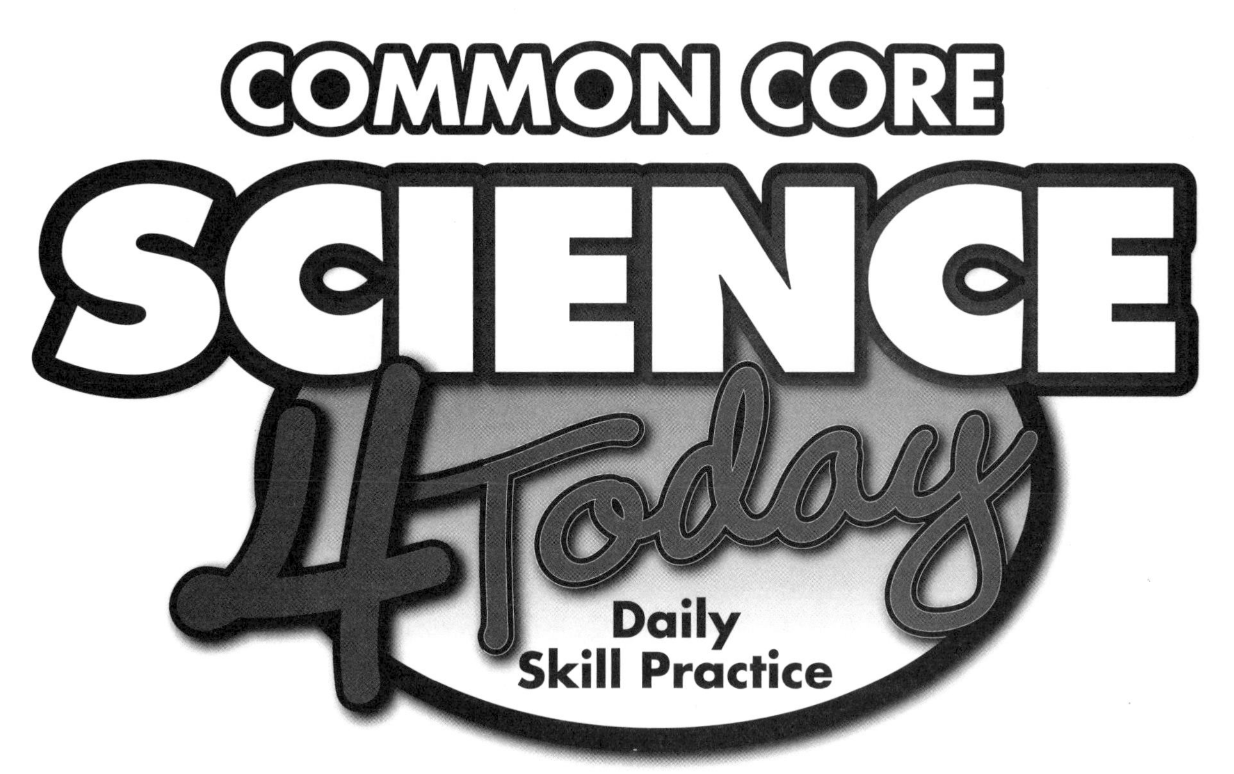

Grade 3

Carson-Dellosa Publishing, LLC
Greensboro, North Carolina

Credits

Content Editor: Angela Triplett
Copy Editor: Karen Seberg

Visit *carsondellosa.com* for correlations to Common Core, state, national, and Canadian provincial standards.

Carson-Dellosa Publishing, LLC
PO Box 35665
Greensboro, NC 27425 USA
carsondellosa.com

ISBN 978-1-4838-1126-0
01-135141151

Table of Contents

Introduction

Common Core Science 4 Today is a perfect supplement to any classroom science curriculum. Students' science skills will grow as they support their knowledge of science topics with a variety of engaging activities.

This book covers 40 weeks of daily practice. You may choose to work on the topics in the order presented or pick the topic that best reinforces your science curriculum for that week. During the course of four days, students take about 10 minutes to complete questions and activities focused on a science topic. On the fifth day, students complete a short assessment on the topic.

Various skills and concepts in math and English language arts are reinforced throughout the book through activities that align to the Common Core State Standards. Due to the nature of the Speaking and Listening standards, classroom time constraints, and the format of the book, students may be asked to record verbal responses. You may wish to have students share their answers as time allows. To view these standards, please see the Common Core State Standards Alignment Matrix on pages 5–8.

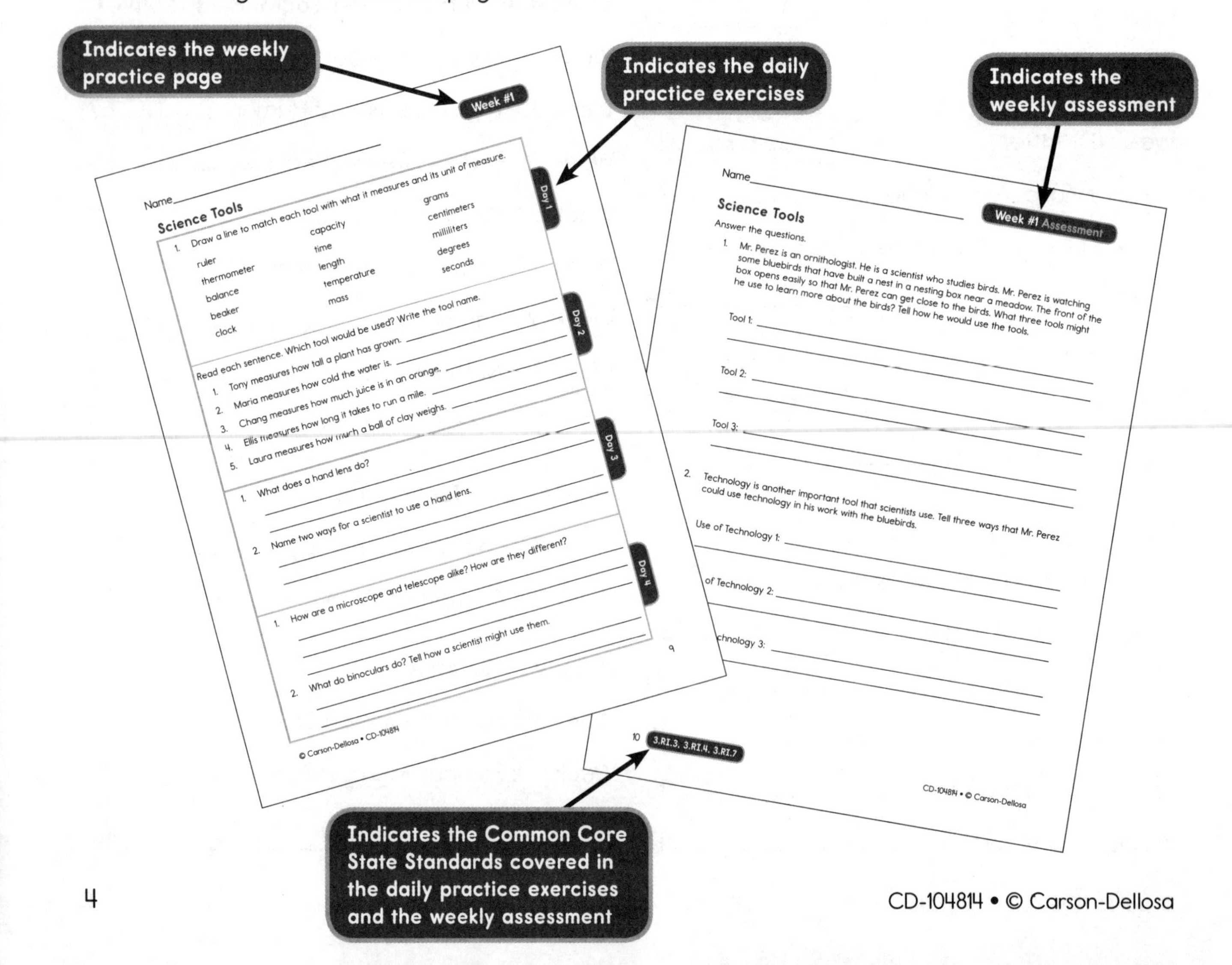

Common Core State Standards Alignment Matrix

English Language Arts

STANDARD	W1	W2	W3	W4	W5	W6	W7	W8	W9	W10	W11	W12	W13	W14	W15	W16	W17	W18	W19	W20
3.RI.1					●		●			●				●	●					
3.RI.2																		●		
3.RI.3	●	●		●							●	●			●		●			
3.RI.4	●		●			●		●	●	●		●		●	●					
3.RI.5					●															
3.RI.6																				
3.RI.7	●		●																	●
3.RI.8		●		●		●		●										●		
3.RI.9																				
3.RI.10				●	●		●							●						
3.W.1							●									●				
3.W.2													●	●			●			
3.W.3																				
3.W.4																				
3.W.5																				
3.W.6																				
3.W.7																				
3.W.8																				
3.W.9																				
3.W.10																				
3.SL.1						●				●				●						
3.SL.2																				
3.SL.3																				
3.SL.4						●					●	●						●	●	
3.SL.5													●						●	
3.SL.6																				
3.L.4								●	●	●	●				●	●	●		●	●
3.L.6												●				●				

W = Week

Common Core State Standards Alignment Matrix

English Language Arts

STANDARD	W21	W22	W23	W24	W25	W26	W27	W28	W29	W30	W31	W32	W33	W34	W35	W36	W37	W38	W39	W40
3.RI.1			●		●	●	●	●	●	●			●				●		●	●
3.RI.2										●			●				●		●	●
3.RI.3	●	●	●	●	●	●	●	●	●	●	●	●	●	●	●	●	●	●	●	●
3.RI.4	●			●									●	●	●			●	●	
3.RI.5																				
3.RI.6										●		●								
3.RI.7					●	●	●	●	●		●									
3.RI.8																	●			●
3.RI.9																				
3.RI.10										●									●	●
3.W.1	●																●			
3.W.2			●							●								●		
3.W.3																				
3.W.4																				
3.W.5																				
3.W.6																				
3.W.7																				
3.W.8																				
3.W.9																				
3.W.10																				
3.SL.1																				
3.SL.2																				
3.SL.3																				
3.SL.4				●		●		●		●	●			●	●	●				
3.SL.5																				
3.SL.6																				
3.L.4	●	●	●	●	●		●		●		●	●	●		●			●	●	
3.L.6		●					●	●	●		●	●								

W = Week

Common Core State Standards Alignment Matrix

Math

STANDARD	W1	W2	W3	W4	W5	W6	W7	W8	W9	W10	W11	W12	W13	W14	W15	W16	W17	W18	W19	W20
3.OA.A.1																				
3.OA.A.2																				
3.OA.A.3		●											●							
3.OA.A.4																				
3.OA.B.5																				
3.OA.B.6																				
3.OA.C.7																				
3.OA.C.7																				
3.OA.D.9																				
3.NBT.A.1																				
3.NBT.A.2			●																	
3.NBT.A.3																				
3.NF.A.1					●												●			
3.NF.A.2																				
3.NF.A.3																				
3.MD.A.1								●	●											
3.MD.A.2		●	●														●			
3.MD.B.3					●															●
3.MD.B.4					●															
3.MD.C.5																				
3.MD.C.6																				
3.MD.C.7																				
3.MD.D.8																				
3.G.A.1											●									
3.G.A.2																				

W = Week

Common Core State Standards Alignment Matrix

Math

STANDARD	W21	W22	W23	W24	W25	W26	W27	W28	W29	W30	W31	W32	W33	W34	W35	W36	W37	W38	W39	W40
3.OA.A.1																				
3.OA.A.2																				
3.OA.A.3																				
3.OA.A.4																				
3.OA.B.5																				
3.OA.B.6																				
3.OA.C.7			●		●	●														
3.OA.D.8																				
3.OA.D.9																				
3.NBT.A.1											●									
3.NBT.A.2															●					
3.NBT.A.3																	●			
3.NF.A.1												●								
3.NF.A.2																				
3.NF.A.3																●				
3.MD.A.1																●				
3.MD.A.2							●								●					
3.MD.B.3																				
3.MD.B.4																				
3.MD.C.5																				
3.MD.C.6																				
3.MD.C.7																				
3.MD.D.8																				
3.G.A.1																				
3.G.A.2																				

W = Week

Name____________________________________

Science Tools

1. Draw a line to match each tool with what it measures and its unit of measure.

ruler	capacity	grams
thermometer	time	centimeters
balance	length	milliliters
beaker	temperature	degrees
clock	mass	seconds

Read each sentence. Which tool would be used? Write the tool name.

1. Tony measures how tall a plant has grown. ____________________
2. Maria measures how cold the water is. ____________________
3. Chang measures how much juice is in an orange. ____________________
4. Ellis measures how long it takes to run a mile. ____________________
5. Laura measures how much a ball of clay weighs. ____________________

1. What does a hand lens do?

____

____

2. Name two ways for a scientist to use a hand lens.

____

____

1. How are a microscope and telescope alike? How are they different?

____

____

____

2. What do binoculars do? Tell how a scientist might use them.

____

____

Name____________________

Week #1 Assessment

Science Tools

Answer the questions.

1. Mr. Perez is an ornithologist. He is a scientist who studies birds. Mr. Perez is watching some bluebirds that have built a nest in a nesting box near a meadow. The front of the box opens easily so that Mr. Perez can get close to the birds. What three tools might he use to learn more about the birds? Tell how he would use the tools.

   Tool 1: ____________________

   ____

   Tool 2: ____________________

   ____

   Tool 3: ____________________

   ____

2. Technology is another important tool that scientists use. Tell three ways that Mr. Perez could use technology in his work with the bluebirds.

   Use of Technology 1: ____________________

   ____

   Use of Technology 2: ____________________

   ____

   Use of Technology 3: ____________________

   ____

Name________________________________

Week #2

The Metric System

Day 1

When people in the United States measure, they use the customary system. Length is measured in inches and feet. Weight is measured in pounds. Capacity uses cups, pints, and gallons. Most other people in the world use the metric system. Every measurement is based on a unit of ten. A meter measures length. There are 100 centimeters in a meter. A liter measures capacity. There are 100 centiliters in a liter. Scientists all over the world, even in the United States, use the metric system when they work.

1. Why might all scientists use the metric system?

____

Day 2

1. Complete the chart with the correct metric amounts.

Length	Capacity	Weight
1.0 meter	_______ liter	1.0 gram
_______ centimeters	100.0 centiliters	100.0 centigrams
1000.0 millimeters	1000.0 milliliters	_______ milligrams

Day 3

Write the abbreviation for each metric measurement word. Then, give an example of something you would measure with that unit.

1. liter ____
2. centimeter ____
3. meter ____
4. gram ____
5. kilogram ____

Day 4

In the metric system, temperature is measured in degrees Celsius (°C). Water freezes at 0°C and boils at 100°C. The temperature on a hot summer day would be around 30°C.

What is the temperature? Circle the best estimate.

1. a popsicle 0°C 45°C
2. a cup of hot chocolate 15°C 85°C
3. room temperature 20°C 60°F
4. swimming weather 5°C 35°C

Name____

The Metric System

Look at each metric measurement. Write the name of an object that is about that measurement.

1. 8 centimeters ______________________
2. 1 meter ______________________
3. 1 liter ______________________
4. 100 milliliters ______________________
5. 1 gram ______________________
6. 2 kilograms ______________________

Use a metric ruler. Draw lines to show each length.

7. 30 millimeters

8. 42 millimeters

9. 12 centimeters

10. Gavin helped his father construct a garage door. The height of the door was a total of 300 centimeters. How many meters tall was the garage door? ______________

11. Write the temperature shown on the thermometer. Tell what you would do in this weather.

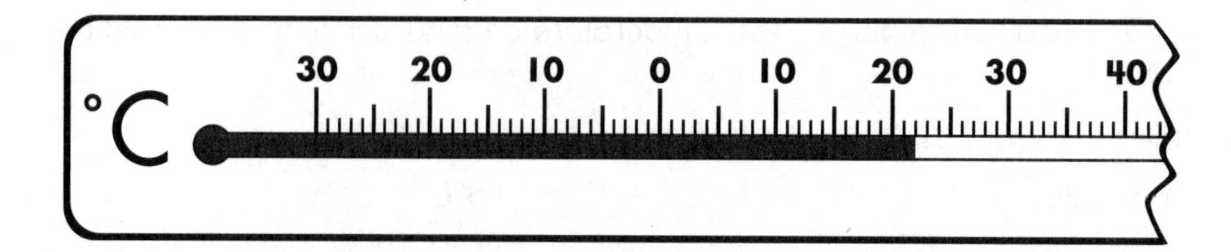

____

____

Name____

Week #3

Mass and Liquid Volume

Day 1

Match each word from the word bank to the correct definition.

balance	graduated cylinder	mass	volume

1. ______________________ a tool used to measure the volume of a liquid
2. ______________________ the amount of matter in an object
3. ______________________ the amount of space that matter occupies
4. ______________________ a tool used to calculate the mass of an object

Day 2

Scientists often measure the volume of liquids using a graduated cylinder. **Graduated** means that the cylinder is marked with measurement units. Scientists always read a graduated cylinder at eye level.

1. What is the volume of liquid in this graduated cylinder? ____________
2. Draw a graduated cylinder with 20 more milliliters of liquid than the one shown.

Day 3

1. Mark has 73 liters of water. Riley has 99 liters of water. How many more liters of water does Riley have than Mark?

____

Day 4

Look at the picture and read the information. What is the mass of the object on the right side of the balance?

1.

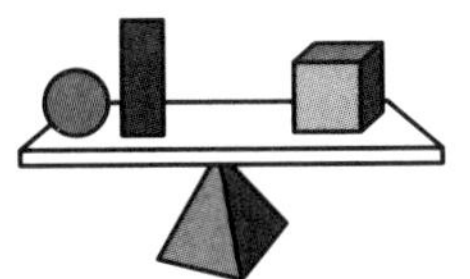

= 5 grams
= 10 grams
=

2.

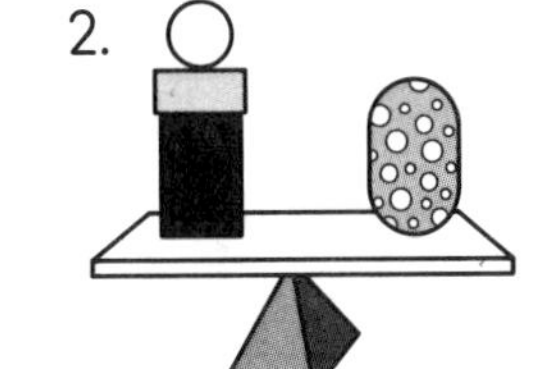

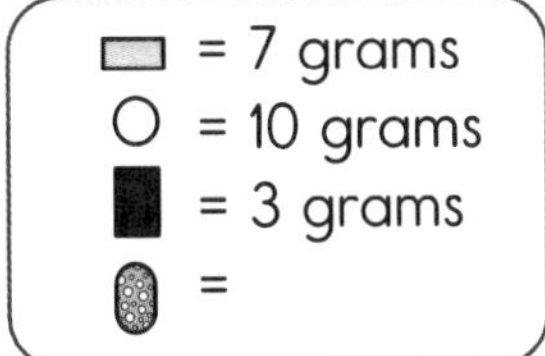

Name____

Mass and Liquid Volume

Answer the questions.

1. Imagine you have two different objects that are the same size. Do the objects have to have the same mass? Explain.

____

____

____

2. Brianna has 3 birdbaths in her yard. Each birdbath holds 3 liters of water. What is the total amount of water in all of her birdbaths? ____________________

3. What is the mass of the objects on the right side of the balance?

A.

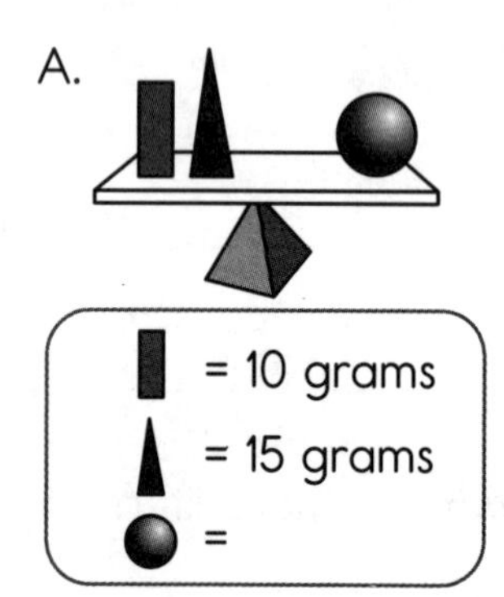

B.

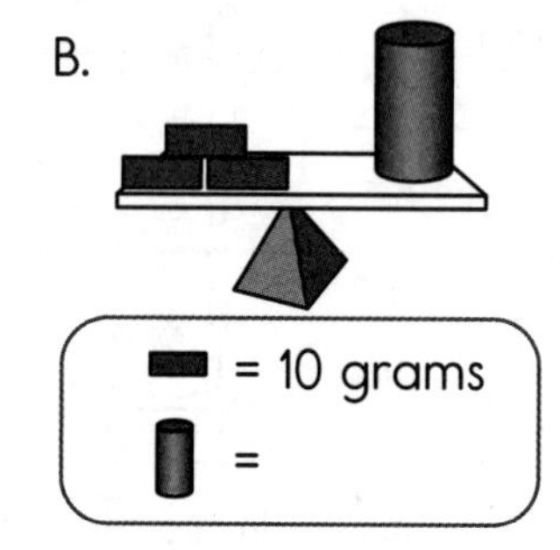

C.

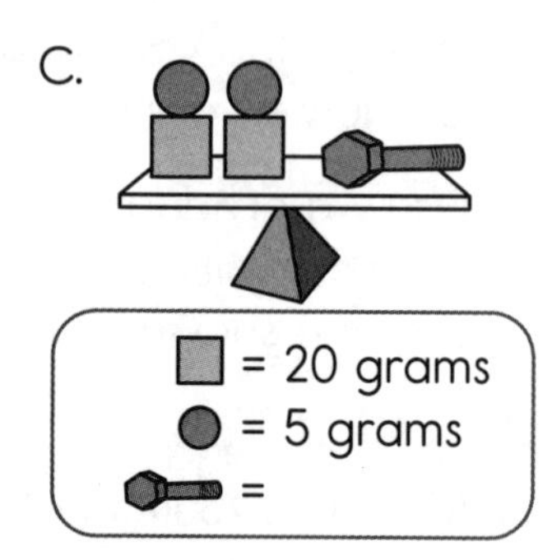

4. Nathan conducted a science experiment using a graduated cylinder. He wanted to find out if oil or water was the heavier liquid. He poured 40 milliliters of water into the cylinder. Next, he poured 20 milliliters of oil into the same cylinder.

   A. What do you think was the result of his experiment? ____________________

   ____

   B. Draw and label a graduated cylinder showing Nathan's experiment.

Circle the best answer.

5. A milk jug can hold about

   A. 2 liters B. 200 liters C. 2,000 liters D. 20,000 liters

Name________________________________

The Science Process Skills

Write a word from the word bank to identify each science process skill.

classify	compare	infer	observe	predict

1. Use what you know to make a guess about what will happen. ________________
2. Use your five senses to learn about the world. ________________
3. Tell how things are alike and different. ________________
4. Sort objects into groups based on characteristics or qualities. ________________
5. Use what you know and what you learn to make conclusions. ________________

1. Scientists sometimes depend on inferences to explain what happened. Look at the pictures. Based on your observations, what can you infer the squirrel is doing?

   ____

1. Scientists must know the difference between facts and opinions. They make conclusions that are based on confirmed observations, or facts. Write three scientific facts.

____

____

____

____

1. Planning and conducting experiments are two more important process skills. The steps must be done in a certain order for the experiment to work. Below is an experiment about digestion. Number the steps from **1** to **5** to show the correct order for the experiment.

   _____ Count to 30 slowly without chewing the cracker.

   _____ Put the cracker in your mouth.

   _____ Use a mirror to see what the cracker looks like in your mouth.

   _____ Draw a picture of the cracker after counting to 30.

   _____ Get a cracker. Draw a picture of it.

Name____________________________________

The Science Process Skills

Read the paragraph. Then, answer the question.

Mario got a good grade on his science test. He decided to put the test on the refrigerator to surprise his mom. He used a magnet to hold the paper on the refrigerator. Mario felt the pull of the magnet as it got close to the steel of the refrigerator. He wondered what else the magnet would stick to. So, he decided to find out. Mario got a dime, a paper clip, a straight pin, scissors, and some aluminum foil. He thought that the dime, pin, and scissors would stick to the magnet, but the others would not. Next, he put the magnet next to each object and watched what happened. If the object stuck to the magnet, he put it in one group. If it did not, he put the item in a different group. He drew a chart to show his results. Mario guessed that the foil and the dime did not have steel in them because they did not stick to the magnet.

1. Name at least four science process skills Mario used. Tell how he used the skills.

Name________________________________

Interpreting Data

Data is a collection of facts from which conclusions can be drawn. Data can be recorded in graphs and charts. Scientists interpret data by studying scientific measurements and observations. They use their conclusions about data to respond to a question. For example, when we want to know what the weather will be for the day, we listen to meteorologists. They use data they have collected and interpreted to make an accurate forecast.

1. Why might a scientist need to interpret data? ________________________________

____

Scientists often measure objects. They record data and display it in a chart or graph. Kayla did an experiment with the weather. It rained for four days. Kayla placed a cup outside. She used a ruler to measure the amount of rain in the cup each night.

1. Look at Kayla's data. Make a chart to show the data.

   Monday, it rained 25 millimeters.

   Tuesday, it rained 38 millimeters.

   Wednesday, it rained 2 millimeters.

   Thursday, it rained 15 millimeters.

1. Make a bar graph to show Kayla's weather data from Day 2.

1. Mr. Foster's class experimented with plants. They planted seedlings and recorded the growth daily for eight days. Graph the class data on a line plot.

Seedlings in Mr. Foster's Class (height in cm)			
4	$4\frac{1}{2}$	$5\frac{1}{2}$	$6\frac{1}{4}$
$4\frac{1}{2}$	$5\frac{1}{2}$	5	$4\frac{1}{2}$
6	$5\frac{1}{4}$	4	$5\frac{1}{2}$
$5\frac{1}{4}$	$6\frac{1}{4}$	6	$5\frac{1}{2}$

Name________________________________

Interpreting Data

Food travels a long distance through the digestive system. Use the following information to create a bar graph. Then, answer the questions below.

Organ	Length (cm)
mouth	7.5 cm
esophagus	25 cm
stomach	25 cm
small intestine	600 cm
large intestine	150 cm

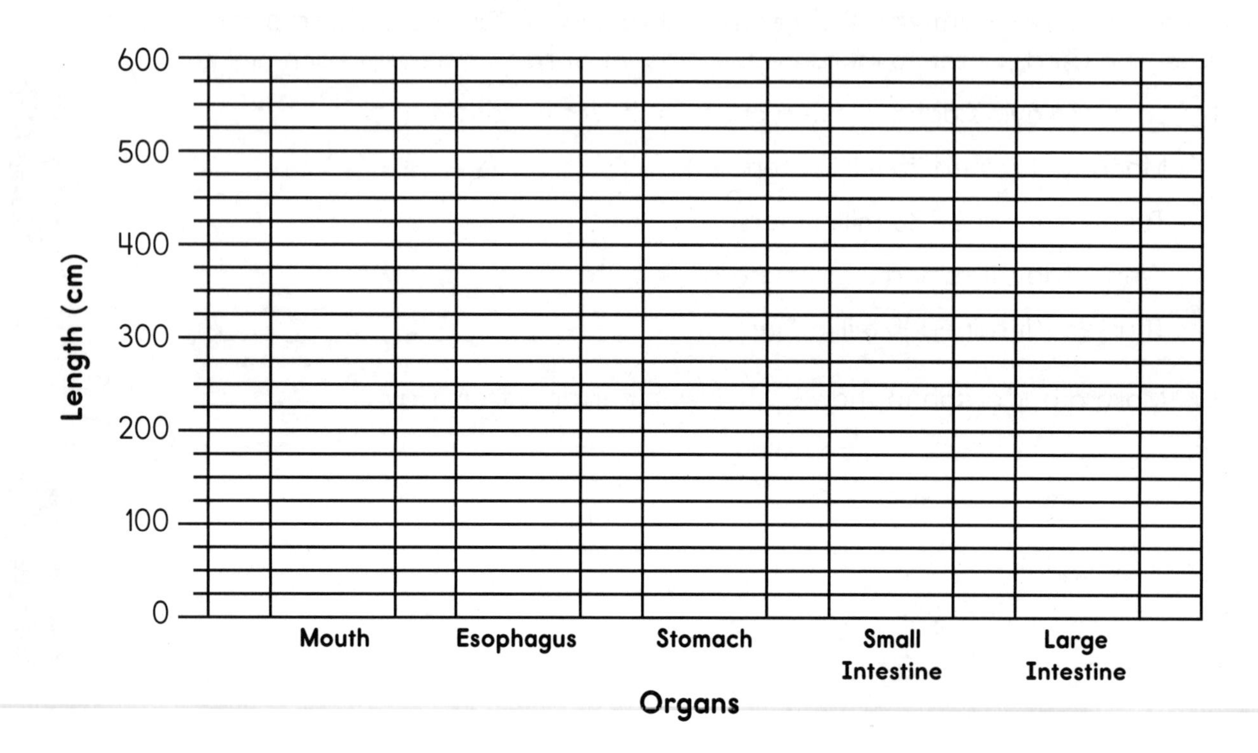

1. What is the total distance that food will travel through the digestive system? ____________________
2. How much longer is the small intestine than the large intestine? ____________________
3. Which two organs are the same length? ________________________________

Name________________________________

Week #6

The Scientific Method

Day 1

To solve a problem, scientists follow steps called the scientific method.

1. Why do scientists use the scientific method?

____

____

2. What might happen if a scientist did not follow the steps of the scientific method in order?

____

Day 2

Write numbers **1** to **7** to show the order of the steps in the scientific method.

_____ observation

_____ comparison

_____ problem

_____ conclusion

_____ hypothesis

_____ presentation

_____ experimentation

Day 3

Write the name of the step in the scientific method being described.

1. Plan and conduct an investigation or activity. ____________________
2. Predict what the results of the investigation will be. ____________________
3. Prepare and share a report that shows the data. ____________________
4. Draw a conclusion from the results of the investigation. ____________________

Day 4

Circle the example in each group that is a good hypothesis.

A. Which melts more quickly—ice cream or ice cubes?

B. Butterflies are pretty.

C. Animals can move seeds.

D. Do worms make the soil better?

A. Maple trees are tall.

B. Plants need sunlight to grow.

C. How much does a cactus grow in one year?

D. I think the rose is the best flower.

Name______________________________

The Scientific Method

Answer the question.

1. Think about an experiment that you have done. Describe what you did and which steps of the scientific method you used. Share with a classmate.

Name___________________________________

A Good Scientist

1. Check the characteristics that you think are important for a scientist to have.

_____ curious _____ patient _____ creative

_____ watchful _____ speedy _____ uninterested

_____ careless _____ eager _____ persistent

2. Choose two characteristics. Tell why they are important qualities for a scientist to have.

____

1. Communication is another important characteristic of a good scientist. Why must a scientist communicate well?

____

____

2. What are two ways that a scientist might communicate?

____

____

Alexander Graham Bell was a scientist. He became interested in sounds when he was a boy. He studied how people talked as he got older. Bell thought of an idea. He wanted to find a way to send sound through a wire. He asked Tom Watson, who knew about electricity, to help make a machine that could do this. The men worked on several devices for a year before finding one that worked. They invented the telephone in 1876.

1. What were two characteristics that made Bell a good scientist? Explain.

____

____

____

Jane Goodall liked chimpanzees. She went to Africa to learn about them. She spent many years living in a tent in the jungle to watch them. She took careful notes about what she saw. Goodall also photographed and filmed the animals in their natural habitats. She found out many new facts about the chimpanzees and the skills they had. Goodall wrote books and gave talks about her discoveries.

1. What were two characteristics that made Goodall a good scientist? Explain.

____

____

____

Name______________________________

A Good Scientist

Answer the questions.

1. Would you make a good scientist? Why or why not? Write a paragraph that explains at least three characteristics that support your opinion.

____

____

____

____

____

____

____

____

____

____

____

2. Choose 3 characteristics that make a good scientist. Give an example of how a scientist might demonstrate each characteristic.

A. ____

____

B. ____

____

C. ____

____

Name____________________________________

Week #8

Matter

Day 1

1. What is matter?

____

____

2. What are the three states of matter? Describe each state.

____

____

Day 2

1. What are atoms?

____

____

Identify each state of matter from the pictures.

2. 

____

3. 

____

4.  

____

Day 3

1. Read each change to matter. Is it a physical or chemical change? Write **P** for a physical change or **C** for a chemical change.

_____ paint	_____ cook	_____ make a mixture
_____ rust	_____ burn	_____ fruit browning
_____ rip	_____ fold	_____ write on

Day 4

1. Describe how water changes states.

____

____

____

2. Alexis wanted to find out how long it would take an ice cube to melt in the hot sun. At 11:06, she placed an ice cube on her sidewalk. At 11:21, the ice cube had completely melted. How long did it take the ice cube to melt?

____

Name________________________________

Matter

Answer the questions.

1. Read each item. Write the name of its state.

   desk ____________________ orange juice ____________________

   air ____________________ water vapor ____________________

   book ____________________ hot chocolate ____________________

2. Why does a chair never change shape?

   ____

   ____

3. Which has more mass—a potato or a potato chip? Explain.

   ____

   ____

4. Think about making cookies. Describe the physical and chemical changes that take place.

   ____

   ____

   ____

5. Why does a shovel rust if it is left outside in the rain?

   ____

   ____

   ____

Name________________________________

Week #9

Force and Motion

Day 1

1. What is a force?

____

2. Describe what kind of force moves a soccer ball. Then, describe the force that stops it.

____

____

____

Day 2

1. What is gravity?

____

2. Why is it harder to walk up the stairs than to walk down them?

____

____

____

____

Day 3

1. Circle the item in each pair that will take more force to move.

A. television
radio

B. book
feather

C. pan
refrigerator

2. Why does it take more force to move the items you circled?

____

____

Day 4

Choose a word from the word bank to complete each sentence.

friction	inertia	lubricant	speed

1. People often slide on ice because there is much less ______________ to stop or slow the motion.
2. A car engine needs ______________ to prevent friction between the parts.
3. An object stays in motion or at rest unless another force acts on it because of ______________.
4. The ______________ of an object is measured by how fast and how far it moves.

Name____

Force and Motion

Answer the questions.

1. How do you know if a force is working?

____

____

2. Imagine that you drop a tennis ball and a bag of potatoes at the same time. Explain what will happen and why.

____

____

____

3. Would it be easier to ride your bike on an icy road or a cement driveway? Explain.

____

____

____

Circle the best answer.

4. Two classes are having a tug-of-war. Mr. Jacob's class is pulling Ms. Ortiz's class closer to the line. What is happening?

   A. Mr. Jacob's class is pulling with less force.

   B. Ms. Ortiz's class is pulling with less force.

   C. Ms. Ortiz's class is pulling with more force.

   D. Both teams are pulling with the same force.

5. Tia and Yuri are racing their bikes to the park. They leave at exactly the same time and follow the same path, but Yuri gets to the park first. Why did Yuri get there first?

____

6. Yuri arrived at the park at 12:14 pm. Tia arrived at 12:27 pm. How much longer did it take Tia to arrive at the park?

____

Name________________________________

Week #10

Simple Machines

Day 1

Use the words in the word bank to write the name of the simple machine represented by each tool shown.

gear	inclined plane	lever	pulley	screw	wedge	wheel and axle

1. ______________ 2. ______________ 3. ______________

4. ______________ 5. ______________ 6. ______________ 7. ______________

Day 2

What kind of simple machine is each object? Write the name of the machine.

1. scissors ______________
2. slide ______________
3. fork ______________
4. window blinds ______________
5. drill ______________
6. fishing rod reel ______________
7. doorstop ______________
8. clothespin ______________

Day 3

1. In the graphic organizer, name three simple machines and describe their uses.

Simple Machine ______________	Simple Machine ______________	Simple Machine ______________
Used for: ______________ ______________ ______________	Used for: ______________ ______________ ______________	Used for: ______________ ______________ ______________

Day 4

1. Beth is opening a can of paint using a screwdriver. Explain two ways that Beth can use the screwdriver to make different simple machines.

____

____

____

____

Name_______________________________________

Simple Machines

Circle the best answer.

1. What simple machine is made when a flat surface is set at an angle to another surface?
   A. a screw
   B. an inclined plane
   C. a pulley
   D. a wheel and axle

2. Which tool is not an example of a lever?
   A. tweezers
   B. nutcracker
   C. hammer
   D. door stopper

Answer the questions.

3. Shannon is pushing on a door, but it does not open. Is she doing work? Explain your answer to a classmate.

   ____

   ____

4. How do simple machines make work easier?

   ____

   ____

   ____

5. Which simple machines are in a wheelbarrow? Tell how each makes work easier.

   ____

   ____

   ____

   ____

Name____

Light

Complete the following sentences to explain some of the properties of light.

1. Light moves in a ____________________ line.
2. When light bounces off an object and changes direction, the light is being ____________________.
3. Light that bends as it moves through different kinds of matter is being ____________________.

Day 1

1. Describe how a rainbow forms in the sky. Explain your answer to a classmate.

____

____

____

____

Day 2

1. What colors make up white light?

____

____

2. Why is an apple red?

____

____

____

Day 3

1. Define transparent. Name one object that is transparent.

____

____

2. Define translucent. Name one object that is translucent.

____

____

3. Define opaque. Name one object that is opaque.

____

____

Day 4

Name______________________________

Light

Answer the questions.

1. What are the attributes of a triangular prism?

____

____

____

2. Imagine that you are using a triangular prism to make a rainbow. You shine the colors on yellow paper. What will you see when you look at the paper? Explain.

____

____

____

3. Imagine that you are using a net to scoop a fish out of a tank. The handle of the net looks like it is broken when you put the net in the water. Why?

____

____

____

4. Why can you see yourself in the surface of a lake?

____

____

Circle the best answer.

5. A mirror is an example of ______________ light.

   A. reflected

   B. refracted

   C. absorbed

   D. translucent

Name________________________________

Heat

Day 1

1. Which are examples of thermal energy? Circle them.

lightning	electricity	snowstorm
wind	burning candle	cooking food
sun	lemonade	fireworks

2. What is heat?

____

____

Day 2

1. Name three examples of thermal energy that you use every day. Tell how each is thermal energy.

____

____

____

____

Day 3

1. Describe how thermal energy moves.

____

____

____

____

Day 4

1. What is a conductor? Name an example of a conductor.

____

____

____

2. What is an insulator? Name an example of an insulator.

____

____

____

Name________________________________

Heat

Answer the questions.

1. Jeff made a hot fudge sundae. Describe how thermal energy moves in it.

____

____

____

2. Daysha needs a warm winter coat. Should she buy one that is wool or cotton? Explain. Then, discuss your answer with a partner.

____

____

____

3. Is a frying pan an insulator or a conductor? Explain.

____

____

____

Write **true** or **false**.

4. _________ Thermal heat always moves to a cooler place.
5. _________ Rubbing your hands together can cause thermal heat.
6. _________ Cold particles bump into each other to make thermal heat.
7. _________ The sun does not make thermal heat.

Name________________________________

Week #13

Electricity

Day 1

What four kinds of energy can electricity make? Write the name of a device that is an example of each.

1. ______________________
2. ______________________
3. ______________________
4. ______________________

Day 2

1. What are the three parts of an atom?

   ____

2. To make an electric charge, what part of an atom is transferred?

   ____

3. What kind of charge must matter have to make an electric charge?

   ____

Day 3

1. If you turn on a light switch and the bulb does not glow, what might you guess about the circuit? Explain.

   ____

   ____

   ____

2. Maria saw 3 strikes of lightning every 4 minutes during a 20-minute electrical storm. How many lightning strikes did she see during the storm?

   ____

Day 4

1. Write three safety rules to follow when using electricity.

   ____

   ____

   ____

   ____

Name____

Electricity

Answer the question.

1. Rap is a spoken-word or music form that means "to talk." Rap's power and beauty comes from the sound and movement of words working with or against a rhythmic background. Raps do not have to rhyme, but they should tell a story or send a message. Create a rap about electricity. Use what you have learned about electricity to write your rap. Then, perform it for your classmates.

Name________________________________

Week #14

Magnetism

For thousands of years, people have noticed that there are some rocks that attract and repel each other. This type of rock is known as a **lodestone**. Because it contains iron oxide, it is naturally **magnetic**. People in ancient China discovered magnetism long before magnets were given a name. They knew that a lodestone swinging on a thread would always point north and that a piece of iron could be **magnetized** by heating it and allowing it to cool while lying north to south. It is thought that the Chinese were also the first people to place a magnetic needle on a pivot so that it could swing freely. This was the first magnetic compass.

1. What is a **lodestone**? ________________________
2. Who is thought to have invented the first magnetic compass? ____________

Day 1

Fill in the blank with a correct word from the word bank.

attract	compass	north	poles	repel

1. A ______________ is a device that has a magnetized needle.
2. The needle in a compass always points ______________.
3. The unlike ends of the magnet ______________ each other.
4. The like ends of the magnet ______________ each other.
5. The ends of magnets are called ______________.

Day 2

1. Draw a compass. Label the parts.

Day 3

1. How is magnetism useful in everyday life?

____

____

____

____

Day 4

Name____________________

Magnetism

Answer the question.

1. Create a new invention that relies on magnetism to work. Draw a diagram and label your invention. Then, explain how it works. Share your invention with your classmates.

Name____________________________________

Living and Nonliving

What does it mean to be alive? Living things come in all shapes, sizes, and colors. You can easily see some living things, such as birds, trees, and people. Other living things, such as mold spores and bacteria, are too small to see without a microscope. All living things have several things in common. For example, they are all made of small units called cells. The cells of living things need energy to work, grow, and repair themselves. All living things also need to create new cells as they grow and develop. Nonliving things, such as rocks and trucks, do not grow or develop.

1. What are the small units that make up all living things? ________________
2. Cells need ________________ to work, grow, and repair themselves.

1. Name four characteristics of living things.

____

____

____

____

____

Match each word from the word bank to the correct definition.

adapt	cells	change	develop	energy	reproduce

1. ______________ tiny parts of living things that perform the basic functions for life
2. ______________ to change because of new conditions
3. ______________ to become different
4. ______________ to make another living thing of the same kind
5. ______________ to grow
6. ______________ the ability to do work

1. Would an icicle be considered living or nonliving? Why?

____

____

2. When does a leaf become a nonliving thing?

____

____

Name____

Living and Nonliving

Some nonliving things can have characteristics of living things. Complete the chart by listing four more characteristics of living things. The first characteristic has been done for you. Then, list five living objects and five nonliving objects. Tree and sun have been listed for you. Decide which characteristics each item has by putting an **X** in the correct column.

Characteristics of Living Things

	Object	energy				
Living	tree	X				
Nonliving	sun	X				

Name____

Week #16

Heredity and Diversity

Day 1

Match each word from the word bank to the correct definition.

diversity	heredity	inherit	species	trait

1. ________________ a feature or characteristic received from a parent
2. ________________ a group of animals that can reproduce
3. ________________ to get from a parent or ancestor
4. ________________ the condition of being different
5. ________________ the passing of characteristics from one generation to the next

Day 2

Kittens look similar to the adult cat when they are young. But, kittens in the same litter can look different. They may be different colors. Some have long fur and some have short fur. Think of other species of animals.

1. What are two advantages of animals in a species being different?

____

____

2. What are two disadvantages of animals in a species being different?

____

____

Day 3

1. Why is it important that animals inherit traits from their parents?

____

____

____

2. What are three specific traits that you inherited from your parents?

____

____

Day 4

What hereditary traits do the animals below have that help them in their environments?

1. snakes ____

____

2. polar bears ____

____

3. camels ____

____

4. fish ____

____

Name______________________________

Heredity and Diversity

Answer the questions.

1. How does heredity affect a species?

____

____

____

2. Name two differences in traits that might occur in animals of the same species. Give an example of each.

____

____

____

____

____

____

3. Do you think it is better to have diversity in a species? Give reasons to support your opinion.

____

____

____

____

____

____

____

Name_______________________

Week #17

Plants

Day 1

1. What four things do plants need?

____

____

2. Ella bought a plant to keep in her bedroom. She watered it every day with a half liter of water. After 9 days, how many liters of water had she used?

____

Day 2

1. Label the diagram to show the parts of a plant. Then, tell what each part does.

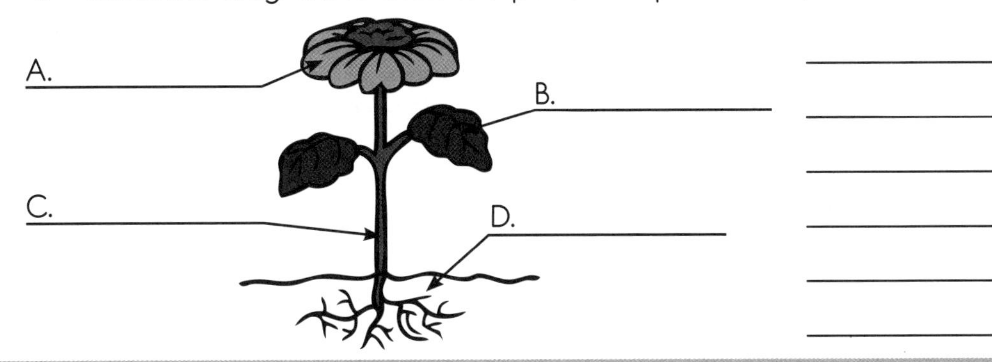

____

____

____

____

____

____

Day 3

1. What is a seed?

____

____

2. Tell at least two ways that seeds are moved.

____

____

____

Day 4

Write **true** or **false**.

1. _________ Photosynthesis is the process in which plants make their own food.
2. _________ Chlorophyll is the energy source that helps plants make food.
3. _________ Water and carbon dioxide join together inside the leaves.
4. _________ Sugar is the food plants make.
5. _________ Plants give off carbon dioxide as a waste.

Rewrite any false statements to make them true.

____

____

Name________________________________

Plants

Answer the questions.

1. Anna did an experiment. She cut a one-inch circle out of construction paper and attached it to the top of a leaf with a paper clip. After one week, she removed the paper circle. What did she see? Explain what happened.

2. What do plants need to make food? What is this process called?

3. Describe three ways that plants are important to people.

Name____

Week #18

Animals

Day 1

1. Name four things all animals need.

____

2. Choose one animal. Tell how your animal meets its needs. Describe your animal and its needs to a classmate.

____

____

Day 2

1. Animals have body parts that help them get food. Name three animals. Tell how their body parts help them get food.

Animal 1: ____

____

Animal 2: ____

____

Animal 3: ____

____

Day 3

1. What are three ways that animals protect themselves against predators? Name one animal that uses each method.

____

____

____

____

____

Day 4

Use the words from the word bank to tell how different animals adapt to winter.

adapt	arctic hare	bear	goose	hibernate	migrate

1. Some animals ______________ to places that have warm weather in the winter. One such animal is the ______________.
2. Some animals ______________ to the cold weather by eating different foods or having body parts that change. The ______________ is one of these animals.
3. Another group of animals ______________ and sleep through the winter. A ______________ does this.

Name________________________________

Animals

Answer the questions.

1. How do fish get oxygen?

____

____

2. The desert kangaroo rat lives in the desert. How does it get water?

____

____

3. Name two reasons that animals need shelter.

____

____

4. Describe how a hawk uses two of its body parts to get food.

____

____

____

5. List three ways that humans can harm animal habitats.

____

____

____

____

____

____

Name________________________________

Animal Groups

1. There are two main groups of animals in the animal kingdom. What are they?
2. To which of the two main groups of animals do insects belong? Explain.
3. Choose two insects. Tell two ways they are alike and two ways they are different.

1. What are two characteristics of reptiles? Name two reptiles.
2. What are two characteristics of amphibians? Name two amphibians.

1. What are two characteristics of birds? Name two birds.
2. What are two characteristics of fish? Name two fish.

1. What are the four characteristics of mammals?
2. To which group of animals does a whale belong? Explain.

Name____________________

Animal Groups

Answer the questions.

1. What are the two main groups of animals? Explain the difference.

2. Look at this animal. To which group does it belong? How do you know?

3. To which group does a bat belong? Explain how you know to a classmate.

4. A butterfly and a bird both have wings, and they come from eggs. Do they belong to the same group of animals? Why or why not?

5. Choose an animal. Use what you know about that animal's characteristics to write a poem. Share your poem with your class.

Name____

Week #20

Life Cycles

Day 1

1. What is a life cycle?

____

2. Do all living things look like the adult when they begin life? Explain.

____

____

____

Day 2

Use the words in the word bank to complete the sentences about how a plant grows.

cones	flowers	leaves	life cycle	root	seed	seedling	stem

1. All plants start from a ______________.
2. First, the ______________ begins to grow.
3. Then, the seed opens and a ______________ begins to grow underground.
4. Soon, the ______________ and ______________ grow above ground.
5. The plant grows to look like the adult and will grow ______________ or ______________ that hold the seeds.
6. Finally, the seeds fall to the ground and the ______________ begins again.

Day 3

1. What is metamorphosis?

____

____

2. Write numbers **1** through **5** to show the life cycle of the frog.

_____ Tadpoles hatch, and can swim in the water and breathe with gills.

_____ The tail of the tadpole disappears.

_____ An egg, covered in a jellylike material, is laid in the water.

_____ A frog hops out of the water to dry land.

_____ Lungs and legs grow on the tadpole.

Day 4

1. Use the following data about a butterfly's life cycle to construct a bar graph.

Stage: egg—4 days
larva—14 days
chrysalis—10 days
butterfly—14 days

Name____

Life Cycles

Answer the questions.

1. Label the diagram to show the life cycle of a butterfly.

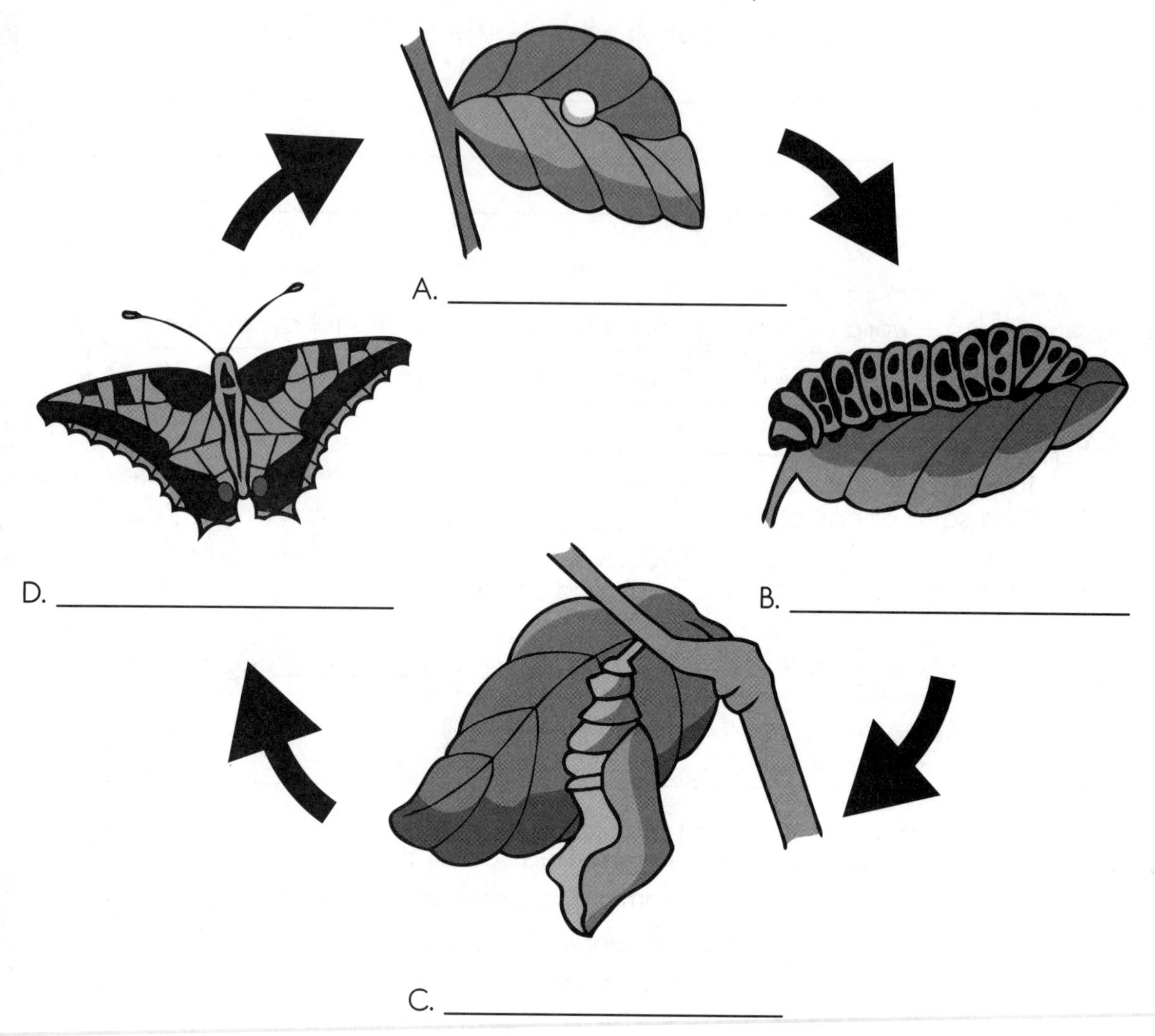

2. Describe the life cycle of a butterfly.

____

____

____

____

____

____

Name_______________________________

Week #21

Ecosystems

Day 1

Draw a line to match each word to its meaning.

1. ecosystem	the place an animal lives where all of its needs are met
2. habitat	all of the groups of living things living in a place
3. population	all of the living and nonliving things in a place
4. environment	a group of one kind of living thing living in a place
5. community	everything that is around a living thing

6. Name 4 things that are a part of your classroom environment.

____

Day 2

1. Name five living things in a forest ecosystem.

_______________ _______________ _______________

_______________ _______________

2. Name five nonliving things in a forest ecosystem.

_______________ _______________ _______________

_______________ _______________

Day 3

Write the name of the ecosystem that correctly completes each sentence.

1. A _______________ ecosystem is very dry.
2. It rains almost every day in a _______________ _______________ ecosystem.
3. Ice and snow cover the land most of the year in the _______________ ecosystem.
4. Coral reefs grow and colorful fish swim in a warm, salty _______________ ecosystem.
5. Water lilies, frogs, and turtles live in a _______________ ecosystem.

Day 4

1. What are three changes, natural or human made, that might happen in an ecosystem?

____

____

____

2. In your opinion, which of these changes would be the most harmful to an ecosystem?

____

____

Name______________________________

Ecosystems

Answer the question.

1. Choose one ecosystem. Draw a picture of it. Include and label at least 10 living or nonliving things that are found in that ecosystem.

Name________________________________

Week #22

Food Chains

Day 1

1. What is a producer?

____

2. Where does a producer get its energy?

____

3. Why are plants important producers?

____

Day 2

1. Describe the three kinds of consumers. Give an example of each.

____

____

____

____

2. Where does a consumer get its energy?

____

Day 3

1. What is a decomposer? Give an example of two decomposers.

____

____

2. Why are decomposers important in an environment?

____

____

Day 4

1. What is a food chain?

____

____

2. Write **1** to **4** to show the order of consumers and producers in a pond food chain.

_____ fish _____ plant _____ duck _____ insect

Name____

Food Chains

Answer the questions.

1. Write **P** if the organism is a producer. Write **C** if the organism is a consumer. Write **D** if the organism is a decomposer.

_____ insect
_____ rose
_____ worm
_____ lettuce
_____ mushroom
_____ bear
_____ human
_____ tree
_____ bacteria

2. Describe the food chain below. Use the words in the word bank in your description.

consumer	energy	producer

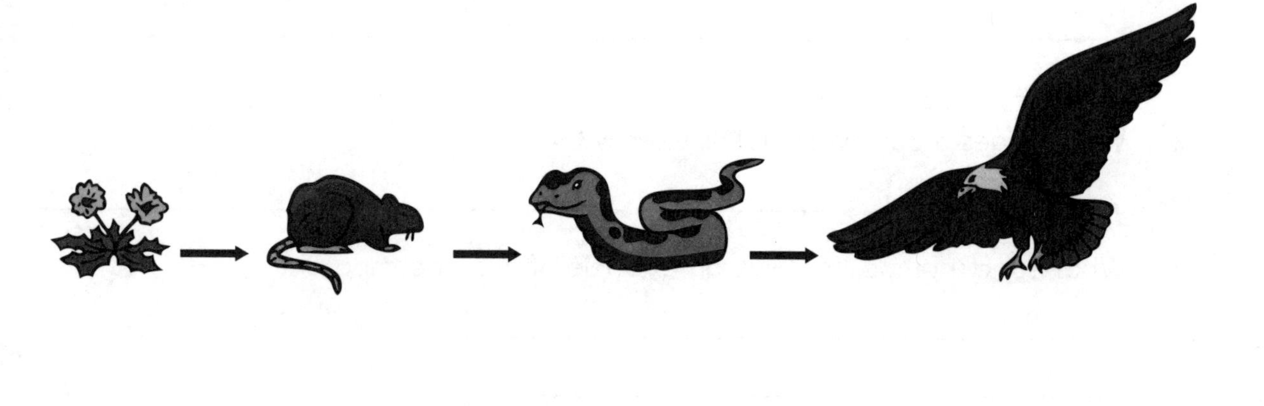

____

____

____

____

Circle the best answer.

3. What is the most important source of energy?

A. sunlight

B. bacteria

C. plants

D. animals

Name________________________________ **Week #23**

Rocks and Minerals

Day 1

1. What is a mineral?

    ____

2. What are three properties of minerals?

    ____

    ____

3. What is the relationship between rocks and minerals?

    ____

Day 2

1. What are the three kinds of rocks? Tell how each type is formed.

    Type 1: ______________________________

    ____

    Type 2: ______________________________

    ____

    Type 3: ______________________________

    ____

Day 3

Read each sentence. Tell what kind of rock each hiker found. Explain how you know.

1. Leslie is hiking in the desert. She picks up a rock. It breaks apart in her hand.

    ____

2. Travis is walking along a trail through the mountains. He sees a place where some rocks have broken off. There are layers of a hard blue rock showing.

    ____

3. David is in Hawaii hiking along a trail near a volcano. He sees many black rocks filled with holes.

    ____

Day 4

Write **true** or **false**.

1. __________ The rocks on Earth are always changing.
2. __________ Wind and water break old rocks down, which become sedimentary rock.
3. __________ Magma cools and forms metamorphic rocks.
4. __________ Sedimentary rocks can be made into metamorphic rocks.
5. __________ Heat and pressure help form igneous rocks.

Name________________________________

Rocks and Minerals

Answer the questions.

1. The rock cycle describes how rocks form and change. Write the names of the kinds of rocks to complete the rock cycle.

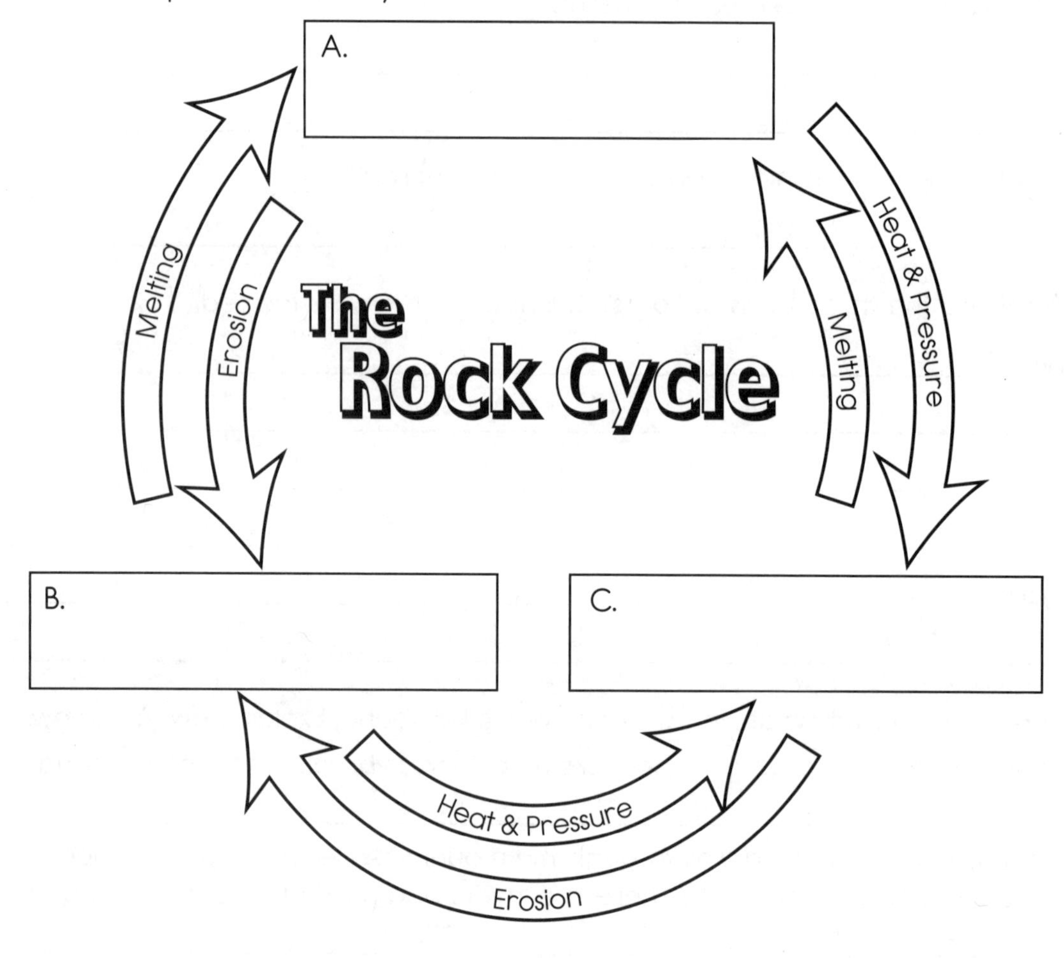

2. Lee has a rock collection containing 27 rocks. She keeps them on shelves in a bookcase in her room. There is one shelf for each type of rock. Each shelf has the same number of rocks on it. How many rocks are on each shelf?

____

____

3. A sculptor is going to carve a statue. She is looking at limestone and marble. Which will she most likely choose? Explain.

____

____

____

Name________________________________

Week #24

Fossils

Day 1

Draw a line to match each word with its meaning.

1. mold
2. paleontology
3. fossil
4. cast
5. dinosaur

the shape of a living thing left as an imprint

a kind of animal that lived long ago

the shape of a living thing made when mud or minerals fill a space

the study of fossils

something left over from a living thing that died long ago

Day 2

1. Write numbers **1** through **5** to show how a fossil is made.

_____ The soft parts rot.

_____ Layers of small rocks, sand, and mud cover the organism.

_____ The plant or animal dies.

_____ A print of the organism remains in the rock.

_____ The pressure of the layers of sediment forms rock.

Day 3

1. How do we know that dinosaurs lived on Earth?

____

____

2. Why must a paleontologist work carefully when digging up fossils?

____

____

Day 4

1. Professor Ross has two fossil teeth in his classroom. One is long and pointed. The other is wide and flat. List two facts that Professor Ross can state about these teeth. Explain the basis for each fact.

Fact 1: _______________________________________

____

____

Fact 2: _______________________________________

____

____

Name____________________________________

Fossils

Answer the questions.

1. Why are most fossils found in sedimentary rocks?

____

____

2. What is the difference between a mold and a cast fossil?

____

____

____

3. How might knowing about plants and animals living today help scientists learn about organisms that lived long ago?

____

____

____

4. Identify three tools a paleontologist needs to help dig fossils and explain how they are used. Share your answers with a partner.

Tool 1: ____

____

Tool 2: ____

____

Tool 3: ____

____

Name____________________________________

Week #25

Land Changes

Draw a line from each landform to its meaning.

1.	mountain	flat land rising above the surrounding land
2.	valley	a wide, flat area of land
3.	canyon	a low place between mountains
4.	plain	a piece of land totally surrounded by water
5.	plateau	a deep valley with steep, high sides
6.	island	a very high, pointed piece of land

Day 1

1. What is weathering?

____

____

2. What are three weather-related forces that cause weathering?

____

3. How do plants cause weathering?

____

____

Day 2

1. How is erosion different from weathering?

____

____

2. Describe two forces that cause erosion.

Force 1: ____________________________________

____

Force 2: ____________________________________

____

Day 3

Read about some changes to the land. Write the name of the event that caused each.

1. The top of the mountain was gone. Red lava poured out of it. Smoke and ash filled the sky and drifted in the wind. The whole forest was on fire.

____

2. The ground began to shake. Suddenly, a large crack formed in the ground. In a nearby house, the windows broke, and the building moved off of its cement pad.

____

3. It had rained so much that water flowed over the banks of the river and ran through the town. The force of the rushing water washed away many things.

____

Day 4

Name______________________________________

Land Changes

Answer the questions.

1. A farmer is plowing his fields. Which picture shows the best way to plow the soil to prevent erosion? Explain.

A. 

B. 

____

____

2. A farmer plowed his fields. He plowed 9 vertical rows in three fields and 8 horizontal rows in four fields. How many rows did he plow altogether? ____________________

Circle the best answer.

3. What is the process where soil is moved from one place to another?
   A. eruption
   B. weathering
   C. erosion
   D. orbiting

4. Which of the following can change the ground quickly?
   A. earthquakes
   B. glaciers
   C. hurricanes
   D. growing plants

5. Which is an example of weathering?
   A. gravity causing a mudslide
   B. water moving across the land
   C. lava flowing down a volcano
   D. ice freezing in a rock

Name________________________________

Week #26

Weather

Day 1

Circle the best answer.

1. Which is not a property of weather?

   A. land

   B. temperature

   C. wind

   D. precipitation

2. Which of the following affects the weather the most?

   A. the clouds

   B. the rain

   C. the sun

   D. the air

Day 2

Write words from the word bank to name the clouds.

cirrus	cumulus	cumulonimbus	stratus

1. ______________

2. ______________

3. ______________

4. ______________

Day 3

Tell how each tool helps you understand the weather.

1. thermometer ____
____

2. anemometer ____
____

3. rain gauge ____
____

4. weather map ____
____

Day 4

1. What is a meteorologist?

____

____

2. How do air masses affect the weather?

____

____

Name____

Weather

Answer the questions.

1. Tell two ways that the weather affects you.

____

____

____

2. During a storm that lasted 21 minutes, Ben saw a lightning bolt about every 3 minutes. How many lightning bolts did Ben see during the storm? ____________

3. Suppose you see some cumulus clouds in the sky. What kind of weather are you having?

____

What activities would you suggest doing in this kind of weather? Compare your answers with a partner.

____

____

Circle the best answer.

4. Which is not a kind of precipitation?

   A. snow

   B. rain

   C. sleet

   D. clouds

5. How does wind move?

   A. from cold areas to hot areas

   B. from hot areas to cold areas

   C. from areas of high pressure to areas of low pressure

   D. from areas of low pressure to areas of high pressure

Name________________________________

Water

Day 1

1. Name five kinds of water features found on Earth.

____

____

2. Why does Earth look mostly blue from space?

____

____

Day 2

1. Ms. Espinosa has a fish tank in her classroom. Does her fish tank hold 100 liters of water or 100 milliliters of water? ____________________

2. Dion filled a bucket with water to mop his floor. Does his bucket hold 8 liters of water or 8 milliliters of water? ____________________

3. Trisha poured some water on her sidewalk on a hot, sunny day. It evaporated in less than 5 minutes. Did she pour 10 liters of water or 10 milliliters of water on her sidewalk? ____________________

Day 3

1. What are the two main kinds of water? Tell where each is found and why each is important.

____

____

____

____

Day 4

Draw a line to match each word to its meaning.

1. water cycle	water when it is a gas
2. evaporation	when heat is removed from a gas, changing it to liquid water
3. condensation	when heat is added to liquid water, changing it to a gas
4. water vapor	the process where water is removed from the surface of Earth and returned back to Earth

Name____________________________________

Water

Look at the diagram. Answer the questions.

1. What does the diagram show?

____

2. Describe what is happening in the diagram.

____

____

____

____

____

____

Circle the best answer.

3. What causes water to evaporate on Earth?

A. the sun B. the lakes

C. the mountains D. the clouds

4. When do clouds form?

A. when water vapor heats B. when water vapor condenses

C. when the air above water evaporates D. when the air above water condenses

Name____________________________________

Week #28

Planets

Day 1

1. Look at the diagram. Label the planets.

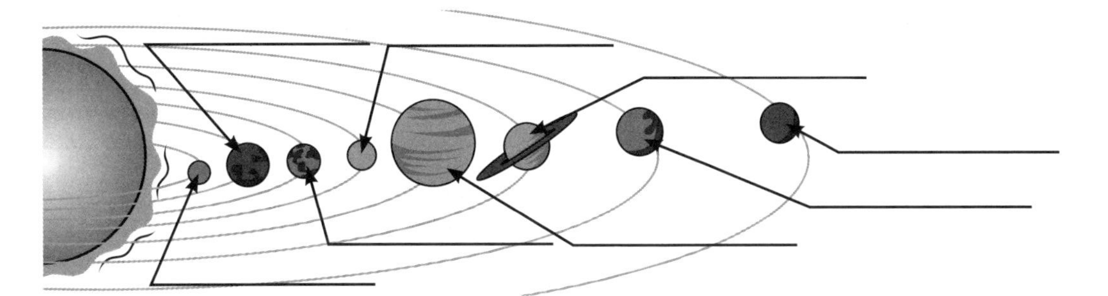

Day 2

Write a word that correctly completes each sentence.

1. The ________________ is the center of the solar system.
2. The ________________ of the sun is the force that holds the planets in place.
3. All of the planets ________________ in a circle around the sun.
4. Earth also spins, or ________________, on its axis.
5. The inner and outer planets are separated by the ________________ belt.

Day 3

Write **true** or **false** for each statement.

1. _________ Mercury moves the fastest around the sun.
2. _________ Uranus is the hottest planet.
3. _________ Earth is the only planet with living things.
4. _________ Saturn is the only planet that has rings.
5. _________ Jupiter is a gas planet.

Day 4

1. What are the names of the inner planets? How are these planets alike?

____

____

____

2. What are the names of the outer planets? How are these planets alike?

____

____

____

Name________________________________

Planets

Answer the questions.

1. Name two materials that Earth has that make it possible for the planet to support life.

____

____

2. Name two ways a planet's distance from the sun affects it.

____

____

____

3. What would happen to Earth if there was no sun? Explain your answer to a partner.

____

____

Circle the best answer.

4. Why is the sun's gravity so strong?

   A. The sun is very hot.

   B. The sun is very bright.

   C. The sun is very big.

   D. The sun is made of gas.

5. Which planet is an outer planet?

   A. Earth

   B. Mercury

   C. Mars

   D. Neptune

Name________________________________

Week #29

Earth and Moon

Write **true** or **false** for each statement.

1. ________ Earth tilts on it axis.
2. ________ The moon makes its own light.
3. ________ The sun is the largest object in the night sky.
4. ________ The moon revolves around Earth.

Rewrite any false statements to make them true.

____

____

Day 1

Write a word that correctly completes each sentence.

1. We get day and night because Earth ______________ on its axis.
2. It takes ______________ hours for Earth to make one rotation.
3. The side of Earth facing away from the sun has ______________.
4. The side of Earth facing toward the sun has ______________.
5. It takes about ______________ days for the moon to orbit Earth one time.

Day 2

1. Why does the moon seem to change shape?

____

____

2. Describe a full moon. Why does it look this way?

____

____

3. Describe a new moon. Why does it look this way?

____

____

Day 3

1. What are two reasons that Earth has seasons?

____

____

2. Look at the diagram. What season is it in the Northern Hemisphere?

____

Day 4

Name______________________________

Earth and Moon

Answer the questions.

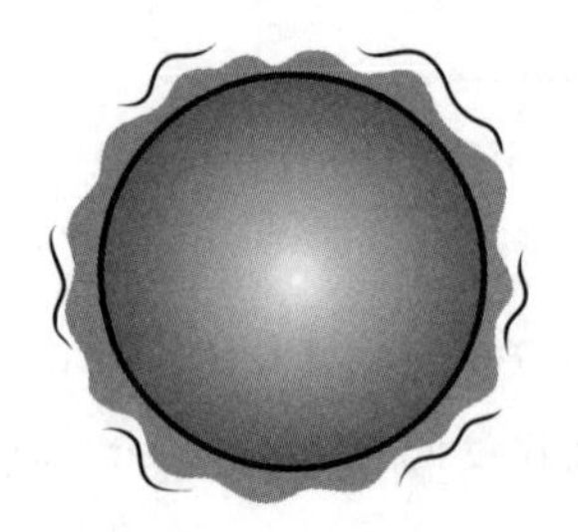

1. Look at the diagram. What season is it in the Northern Hemisphere?

 ____

2. What are two characteristics of the season for the Northern Hemisphere shown above?

 ____

 ____

 ____

3. What would happen if Earth was not tilted?

 ____

Circle the best answer.

4. What are the changes in the moon called?

 A. nights

 B. years

 C. phases

 D. seasons

5. What is the moon called when it looks like a sliver of light?

 A. new moon

 B. full moon

 C. waxing moon

 D. crescent moon

Name________________________________

Week #30

Rain Forests and Humans

Day 1

More than half of all plant and animal species in the world live in the rain forest. Scientists believe that millions more species exist but have not yet been discovered. However, in the last 50 years, nearly half of the rain forests have been destroyed.

1. Why might scientists be concerned about the loss of the rain forests?

____

____

____

____

Day 2

The rain forest is of great interest to many companies. Loggers cut large areas of trees to harvest the wood for a variety of products. Miners strip the land of trees to dig out minerals in the ground. The land is left bare once the companies take what they want.

1. Is it a good idea to leave the land bare? Explain.

____

____

____

____

Day 3

People have lived in the rain forest for thousands of years. Each year, they burn parts of forest so that they can grow crops. So many people use the slash and burn method that huge clouds of smoke are carried to other continents. Moreover, rain forest soil is thin and does not have many nutrients. After farming the land for several years, the soil no longer grows healthy plants. The people need to clear more trees to make new fields.

1. Tell two ways that the slash and burn method is harmful.

____

____

Day 4

The native people use the plants in the rain forest for medicines. Drug companies are working with them to find out which plants they use to cure different illnesses. Then, scientists research how to use the plants to make new medicines. For example, the rosy periwinkle, a flower grown in the Madagascar rain forest, is now used to make a drug that helps cancer patients.

1. How is the relationship between drug companies and native people a positive one?

____

____

Name________________________________

Rain Forests and Humans

Circle the best answer.

1. What is the slash and burn method?

   A. Logging companies dig out minerals.

   B. Workers find and cook food.

   C. Native people clear land to grow crops.

   D. Logging companies cut down trees.

2. Why is it important to work with the native people living in a rain forest?

   A. They know about the plants and animals.

   B. They want to sell their land.

   C. They do not want strangers in the rain forest.

   D. They want to make money logging.

Answer the question.

3. Write a paragraph telling three reasons why the rain forest is important. Share your paragraph with a partner.

____

____

____

____

____

____

____

____

Name____________________________________

Week #31

Reduce, Reuse, Recycle

Day 1

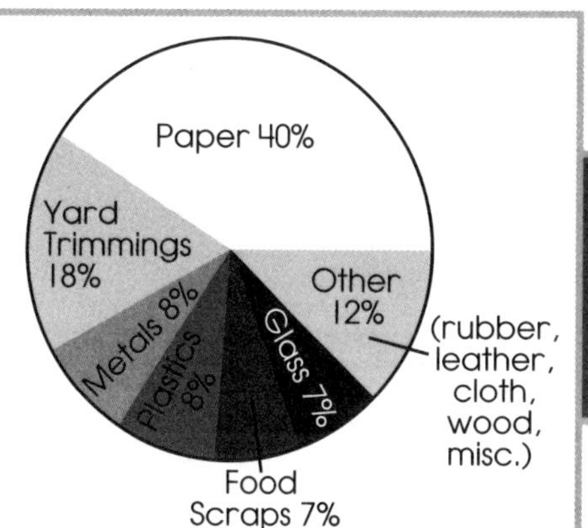

1. Which kind of trash is thrown out the most?

   ____

2. What are three examples of this kind of trash that you throw out?

   ____

3. Which two categories make up about one-half of the pie chart?

   ____

Day 2

Draw a line to match each word with its meaning.

1. waste	to make something new out of something old
2. reduce	to protect something from destruction
3. reuse	to use less of something
4. recycle	to find a new use for something
5. conserve	something that is thrown out

Day 3

1. Kristen has a shoe box. What are two ways that she can reuse it?

   ____

   ____

2. Baytown has a water shortage. What are two ways people can reduce the amount of water they use?

   ____

   ____

   ____

Day 4

Write a word to complete each sentence.

1. Save ______________ by turning off the faucet when you brush your teeth.
2. Save ______________ by turning off lights that are not being used.
3. Give ______________ you have outgrown to someone who can wear them.
4. Use dishes instead of ______________ plates when you eat.
5. Recycle your soda cans because the ______________ can be used to make new products.

Name____________________________________

Reduce, Reuse, Recycle

Circle the best answer.

1. Which product comes from trees?
   A. plastic
   B. glass
   C. paper
   D. aluminum

2. Which is not a product that can be made from recycled glass?
   A. roads
   B. bowls
   C. earrings
   D. shoe box

Answer the questions.

3. Provide an example for each way you can help conserve Earth's resources. Then, share your answers with a partner.

   Reduce: ____
   ____

   Reuse: ____
   ____

   Recycle: ____
   ____

4. Why is it important to reduce, reuse, and recycle items?
   ____
   ____
   ____
   ____
   ____
   ____

Name________________________________

Resources

1. What is a natural resource?

____

____

2. Write **N** beside each item that is a natural resource.

_____ air _____ corn _____ water

_____ cow _____ oil _____ coal

_____ shirt _____ electricity _____ paper

1. Look at each picture. Tell why it is a resource.

1. What is a renewable resource? Give two examples.

____

____

2. What is an inexhaustible resource? Give two examples.

____

____

1. A lumber company cut down 250 trees from a woodland of 500 trees. What fraction of the trees did the lumber company cut from the woodland?

____

2. The lumber company then sold 25 of the trees that were cut down to a paper company. What fraction of the trees was sold to the paper company?

____

Name______________________________________

Week #32 Assessment

Resources

Answer the questions.

1. Look at the picture. How is it a resource?

____

____

____

____

2. Is it important to care for inexhaustible resources? Give an example in your explanation.

____

____

3. How does a lumber company make sure that a tree is a renewable resource?

____

____

Circle the best answer.

4. What kind of resources are plants and animals?

   A. exhaustible

   B. inexhaustible

   C. renewable

   D. nonrenewable

5. Which is not an example of a nonrenewable resource?

   A. coal

   B. iron

   C. oil

   D. water

Name________________________________

Week #33

Animal Conservation

Draw a line to match each word to its meaning.

1.	conservation	there is only a small number of this living thing
2.	extinct	a place where the habitat of a living thing is kept safe
3.	endangered	a group of living things is being kept safe by laws
4.	protected	every one of this kind of living thing has died
5.	refuge	to keep the living things on Earth safe

Day 1

1. Dinosaurs lived long ago. How do we know that these animals existed?

____

____

2. Why do scientists think that dinosaurs became extinct?

____

____

____

Day 2

Bluebirds live in holes they find. The holes may be in trees or in fence posts. People started clearing the land to make roads, buildings, and farms. The bluebird population got smaller. Soon, the bird was added to the endangered list.

1. Think about what animals need. What did the bluebird not have?

____

____

2. What did people do that caused the bluebird population to get smaller?

____

____

Day 3

People worried that bluebirds would become extinct. They began to help the birds. People all over the United States built special boxes that the birds could live in. They nailed the boxes to trees beside meadows. Now, the bluebird population is growing. The bluebirds have been removed from the endangered list in some states.

1. Describe two human activities that helped the bluebirds.

____

____

____

____

Day 4

Name____

Week #33 Assessment

Animal Conservation

Read the passage. Then, answer the questions.

The bald eagle is the national symbol of the United States. At one time, many of these birds flew in the sky. Then, farmers began using chemicals to kill insects. Fish and other small animals ate the insects. The chemical transferred to those animals. When the eagles ate the animals, they, too, had the chemical in their bodies. The eagles began laying eggs with thin shells. When the parent birds sat on the eggs, the eggs cracked. Few baby eagles were born. The bald eagle was added to the endangered list.

Scientists were puzzled. The chemicals were not used near the eagles. They worked to solve the problem. Now, those chemicals are not used. People are working to make sure the eagles have everything they need so that their population will continue to grow.

1. How did people harm the eagles?

____

____

____

2. How did knowing about a food chain help scientists discover the problem with eagles?

____

____

____

3. Why is it important to understand the balance of nature?

____

____

____

____

____

Name________________________________

Week #34

Safety

Day 1

Unscramble the words in parentheses to complete the fire safety rules.

1. Put smoke ________________(srcotdeet) near the bedrooms.
2. Make a plan to ________________(speeca) from every room in the house.
3. Choose a place where everyone should meet ________________(duostie).
4. Keep a fire ________________(stuigehrnixe) on each floor in the house.
5. If your ________________(selocht) catch on fire, you should stop, drop, and roll.

Day 2

Think about thunderstorm safety. Write **true** or **false**.

1. __________ Stay away from glass windows and doors.
2. __________ Stand under a tree if there is lightning.
3. __________ Stay away from water.
4. __________ If you cannot reach shelter, crouch down and tuck your head.
5. __________ Talk on a landline phone with an adult.

Rewrite any false statements to make them true.

____

____

Day 3

1. If you are in a building, what should you do during an earthquake?

____

____

____

2. If you are outside, what should you do during an earthquake?

____

____

____

Day 4

Write words to correctly complete the tornado safety rules.

1. Do not wait until you see the ______________________________to move to a safe place.
2. Move to a room that does not have ______________________________.
3. If you are outside and cannot reach shelter, move to a ______________________________.
4. Listen to weather ______________________________on the TV or radio.
5. Stay put until the danger is ______________________________.

Name____

Week #34 Assessment

Safety

Answer the questions.

1. Why is it important to move away from glass windows and doors during earthquakes and tornadoes?

____

____

2. Why should you not stand under a tree during a thunderstorm?

____

____

3. How do smoke detectors help keep people safe during fires?

____

____

4. Why is it important to learn and practice safety rules?

____

____

5. The number 911 is an important number to know. What is it used for? Identify a situation when you would need to call that number. Discuss your answer with a partner.

____

____

____

____

Name________________________________

Nutrition

Day 1

1. What are the five food groups? Give an example of each.

   A. ______________________________

   B. ______________________________

   C. ______________________________

   D. ______________________________

   E. ______________________________

Day 2

1. What did you eat during your last meal? Write the name of each food. Tell which food group it is from.

   ____

   ____

   ____

   ____

Day 3

1. What is a balanced diet?

   ____

   ____

2. Cory's doctor said she should have at least 34 grams of protein every day. Today, she drank a glass of milk containing 8 grams of protein. Cory also ate a cup of yogurt, which contained 11 grams of protein. How many more grams of protein does Cory need today to follow her doctor's orders?

   ____

Day 4

Draw a line to match the food with its primary nutrient.

1.	fish	vitamin A
2.	carrots	vitamin C
3.	milk	iron
4.	spinach	protein
5.	bread	carbohydrate
6.	oranges	calcium

Name____________________________________

Nutrition

Answer the questions.

1. Not all people should eat the same amounts of food. For example, a two-year-old girl will not need to eat as many servings of fruit as an eight-year-old girl. Explain why.

____

____

____

2. Why is it important to eat foods from different food groups?

____

____

3. What happens to a person who eats foods that have too many fats or sugars? Write a short oral report you would give to kindergarteners to explain what would happen.

____

____

____

____

____

____

____

4. Lily is having some friends over after school. What is a nutritious snack that she might share with her friends?

____

Circle the best answer.

5. Which food is part of the grain group?

   A. peas

   B. yogurt

   C. nuts

   D. noodles

Name____________________________________

Week #36

Exercise

Day 1

1. Name two ways that exercise helps your body.

____

____

2. What might happen if you did not exercise your muscles?

____

____

Day 2

1. How does your body change when you do warm-up exercises?

____

____

2. Name two reasons that you should warm up before doing a physical activity.

____

____

____

Day 3

1. Wyatt attended football practice for 1 hour. The team spent $\frac{1}{4}$ hour throwing the ball, and $\frac{1}{2}$ hour running drills. For the last $\frac{1}{4}$ hour, Wyatt's team practiced catching the ball. Did the team spend more time throwing or running drills?

____

2. Tara went to the gym to exercise. She warmed up for 9 minutes and then ran on the treadmill for 22 minutes. Tara did sit-ups for 5 minutes. She stretched for 8 minutes when she finished her workout. How long did Tara exercise at the gym?

____

Day 4

1. Why should you give your body time to cool down after exercising?

____

____

2. How does stretching after exercising help your muscles?

____

____

Name____

Exercise

Write **true** or **false**.

1. ________ When you warm up, you should run very fast.
2. ________ Warming up will help prevent muscles from getting hurt.
3. ________ Bouncing while stretching can cause injury.
4. ________ Exercising can help control stress.
5. ________ A person should do 30 minutes of aerobic activity every day.

Answer the questions.

6. Why is it good to do many different activities and exercises?

____

____

7. Jan stretched for 5 minutes. Then, she jumped rope with her friends for 30 minutes. She sat down right away to eat an apple. Could Jan have done something to make her workout better? Explain. Share your answer with a partner.

____

____

____

Circle the best answer.

8. How does aerobic exercise help your lungs?
   A. They hold more air.
   B. You can breathe faster.
   C. You can smell better.
   D. All of the above.

9. How does exercise help your body?
   A. You can play longer.
   B. You are stronger.
   C. You can move more easily.
   D. All of the above.

Name________________________________

Week #37

Energy Technology

Day 1

In some places, the wind is a constant force. Scientists have found a way to convert it into electricity. Giant turbines are built up above the ground. The turbines have blades on them. As the wind blows, it turns the blades and spins the turbine. The turbine powers a generator, which produces electricity.

1. Why are the turbines built high above the ground?

____

2. A turbine's blade turns 20 times per minute. How many times will it turn in 9 minutes? ______________________________

Day 2

Moving water has energy. People long ago used water energy to turn large rock wheels to grind corn and wheat into flour. People still use water energy today. They build large dams to hold water in lakes. When the water is released, it turns large turbines. The turbines power generators, which produce electric energy. The electricity moves through power lines to light and heat buildings.

1. Is it possible for all electricity to be powered by water? Why or why not?

____

____

Day 3

Solar energy is energy that comes from the sun. Some people use solar energy to heat their houses. Flat, black panels on the roof gather the light. The panels have water-filled tubes inside. The water in these tubes gets hot and travels to a heat exchanger, filled with more water. The heat from the water-filled tubes is transferred to the water in the heat exchanger. The hot water moves through a special heating system to warm a house.

1. Why are the panels on the roof black?

____

____

Day 4

In an atom, electrons circle around the nucleus, which is made up of neutrons and protons. The main energy of the atoms is found in the nucleus. To make nuclear energy, scientists break an atom's nucleus, which creates heat. The heat is converted into steam. Steam powers machines that make electricity. Nuclear energy is good because it does not use fossil fuels, like coal. It does not release pollutants into the air, either. However, the fuel used to split the atoms, uranium, is dangerous once it is used. It must be removed and stored away from living things.

1. Even with the dangers of nuclear energy, why might many countries continue to build nuclear plants?

____

Name____

Energy Technology

Answer the questions.

1. Why is wind a good source of energy? Why might it be a bad source of energy?

____

____

____

2. Would a homeowner living in Florida or Alaska be more likely to use solar power for heating? Explain.

____

____

____

3. Other than energy, what are the benefits of a town building a dam?

____

____

____

____

4. Which form of energy do you think is the best for people to use? Give two reasons in your explanation.

____

____

____

____

____

Name________________________________

Week #38

Computer Technology

Day 1

1. What is a computer?

____

____

2. An abacus was a tool that merchants used to calculate numbers in ancient China. Many people think it was the first computer. Why?

____

____

Day 2

Draw a line to match each word with its meaning.

1.	monitor	a program that makes the computer work
2.	keyboard	the network that links computers all over the world
3.	microprocessor	the part on which people input the information
4.	file	the part that does the computing
5.	Internet	the unit in which information is stored
6.	software	the part where people see the information
7.	hard drive	the part that stores all the information

Day 3

1. Name four ways that people use computer technology to communicate.

____

____

____

____

Day 4

1. Why is a digital camera a type of computer technology?

____

____

____

____

Name______________________________________

Computer Technology

Answer the question.

1. How has computer technology changed society? How do you use computer technology in your everyday life? Use the words from the word bank in your explanation.

communicate	learn	play	work

____

____

____

____

____

____

____

____

____

____

____

____

____

____

____

____

____

____

____

____

____

____

____

Name________________________________

Week #39

Famous Women Scientists

Marie Curie is one of the most famous female scientists. She won many awards, including winning the Nobel Peace Prize twice. She was a chemist and physicist who studied radioactive minerals, nonliving things in nature that gave off energy in the form of rays. Through her work, Curie discovered two new elements and developed the use of X-rays.

1. Why do you think Curie is famous?

____

____

Day 1

As a child, Mae Jemison liked science and math. She learned about medicine in college. When she graduated, she wanted to help people. She worked in several countries that did not have good health care. Jemison returned to the United States and became an astronaut. She was the first African American woman to travel into space. When Jemison left the space program, she continued to help people in other countries by setting up a satellite system that improved health care.

1. How could a satellite system help people in faraway places?

____

____

Day 2

Williamina Fleming never went to school. She began working for an astronomy professor as a maid. The professor saw that Fleming was observant and wise. He asked Fleming to work for him. Fleming watched and listened to the professor. She began to study the stars. Fleming created a system to organize, or classify, stars. In one year alone, Fleming discovered 222 stars! Fleming also identified a white star, which, from its color, showed that the star was about to die.

1. Why is having a system to classify stars important?

____

____

Day 3

You probably know the name Beatrix Potter. She wrote and illustrated *The Tale of Peter Rabbit*. However, did you know that Potter was also a botanist, or plant scientist? She studied fungi, living things that live on dead plant and animal matter. She collected samples of many different fungi and then cut them open to look inside. Then, she would paint pictures filled with details to show what they looked like. Amazingly, Potter made over 300 pictures of mushrooms alone.

1. How might Potter's pictures help other scientists?

____

____

Day 4

Name____________________________________

Week #39 Assessment

Famous Women Scientists

Draw a line to match each word with its meaning.

1. astronomer	a scientist who studies plants
2. physicist	a scientist who studies matter and energy
3. chemist	a scientist who studies stars
4. botanist	a scientist who studies materials and how they work together when joined

Answer the questions.

5. How might an outside interest help a scientist? Give an example.

____

____

____

6. How do many different people from around the world benefit from work scientists do? Give an example.

____

____

____

____

Circle the best answer.

7. Which activities was Beatrix Potter interested in?
   A. botany and writing
   B. chemistry and botany
   C. astronomy and writing
   D. physics and chemistry

8. Why is Mae Jemison famous?
   A. She won two Pulitzer Prizes.
   B. She created a system to classify stars.
   C. She was the first African-American woman in space.
   D. She discovered the X-ray.

Name______________________________

Famous Men Scientists

What do shaving cream, mayonnaise, peanut butter, and cherry punch have in common? George Washington Carver! He was an inventor and scientist who worked with plants. By the 1900s, the soil in the South did not have many nutrients, and it did not hold water well. The cotton plants that had been grown for years had hurt the soil. Carver found that planting peanuts and sweet potatoes helped the soil. To make peanuts a worthwhile crop, he figured out how to make many products out of peanuts, such as shaving cream.

1. How did planting peanuts and sweet potatoes help the soil?

____

Day 1

Thomas Edison had over 1,000 patents with his name on them. A patent gives a person or company with an idea the right to be the only one who can use, make, or sell things with that idea. Edison spent many years learning about electricity and sound. One of his most well-known patents was received for the light bulb. It made using lights in a house safe and useful. It was not too expensive, either.

1. Why would a scientist want to patent an idea?

____

____

____

Day 2

Putting lights in one house was easy. Yet, Edison needed to find a way to get lights into every house. He needed to invent a system to link electricity to each house and business in a town. With more experiments, Edison invented parallel circuits, an underground conductor network, safety fuses, and on-and-off switches.

1. What problem with lights did Edison want to solve?

____

____

____

Day 3

Thomas Edison may have invented an electric system, but it was Lewis Latimer who made improvements to the light bulb that really made it safe. Edison's light bulb only pushed into a socket. The bulb could easily fall out if the lamp tipped. Latimer invented and patented a bottom that had threads. It could be screwed into the socket. Now, the electric lamp was really safe.

1. How did adding screws to the bottom of the bulb make it safer? Explain.

____

____

____

Day 4

Name____________________________________

Famous Men Scientists

Write **true** or **false**.

1. ________ Thomas Edison invented the light bulb.
2. ________ Sweet potatoes can add nutrients to soil.
3. ________ Peanuts can be used to make shaving cream.
4. ________ Lewis Latimer improved the light bulb.
5. ________ Latimer's bulb was improved using a simple machine.

Answer the questions.

6. How did communication help improve the light bulb?

____

____

____

7. Why did Thomas Edison need to invent a system for lights?

____

____

____

8. George Washington Carver invented over 100 uses for peanuts alone. Why do you think he did this?

____

____

____

____

Answer Key

Page 9
Day 1: 1. ruler, length, centimeters; thermometer, temperature, degrees; balance, mass, grams; beaker, capacity, milliliters; clock, time, seconds; **Day 2:** 1. ruler; 2. thermometer; 3. beaker; 4. clock; 5. balance; **Day 3:** 1. A hand lens makes things look larger than they are. 2. Answers will vary. **Day 4:** 1. Answers will vary but may include they are alike because they both magnify things. They are different because a telescope is used to see things that are far away. A microscope is used to see things that are very small. 2. Answers will vary but may include binoculars have lenses that make distant objects appear closer and larger. Answers will vary but may include to observe wildlife from a distance.

Page 10
1. Answers will vary. Check students' reasoning. 2. Answers will vary but may include: write reports, make charts and graphs, do research, and communicate with other scientists.

Page 11
Day 1: 1. Answers will vary but may include even if scientists cannot understand each other's words, they can understand all of the measurements made by other scientists around the world. **Day 2:** 1. length, 100.0; capacity, 1.0; weight, 1,000.0; **Day 3:** 1. L; 2. cm; 3. m; 4. g; 5. kg; Examples will vary. **Day 4:** 1. 0°C; 2. 85°C; 3. 20°C; 4. 35°C

Page 12
1–6. Answers will vary. 7–9. Check students' drawings. 10. 3 meters; 11. The temperature is 22°C. Answers should include appropriate activities for a warm day.

Page 13
Day 1: 1. graduated cylinder; 2. mass; 3. volume; 4. balance; **Day 2:** 1. 50 mL; 2. Check students' drawings to see if they have drawn an accurately marked graduated cylinder and marked it to the 70 mL line. **Day 3:** 26 liters; **Day 4:** 1. 15 grams; 2. 20 grams

Page 14
1. No. One object may have more matter in it, so it will be heavier. 2. 9 L; 3. A. 25 grams; B. 30 grams; C. 50 grams; 4. A. Answers will vary. B. Check students' drawings to see if they have correctly drawn and numbered a graduated cylinder with 40 mL of water on the bottom and 20 mL of oil on the top for a total of 60 mL of liquid. 5. A

Page 15
Day 1: 1. predict; 2. observe; 3. compare; 4. classify; 5. infer; **Day 2:** 1. Answers may vary but should include that the squirrel is gathering and storing nuts for the winter. **Day 3:** 1. Answers will vary. **Day 4:** 1. 3, 2, 4, 5, 1

Page 16
1. Mario **planned** and **conducted** an experiment to see what items would stick to a magnet. He **predicted** which objects would stick to the magnet and **observed** what happened when the magnet got near each item. He **classified** the items as being magnetic or not and **displayed** the data in a chart. Mario **inferred** that some items were not made of steel because they did not stick to the magnet.

Page 17
Day 1: 1. Answers will vary but may include to draw conclusions, classify data, and display data. **Day 2:** 1. Check students' charts. **Day 3:** 1. Check students' graphs. **Day 4:** 1. Check students' line plots.

Page 18
Check students' graphs. 1. 807.5 cm; 2. 450 cm; 3. esophagus and stomach

Page 19
Day 1: 1. Answers will vary but may include it is an organized way to make sure that a problem is solved fully and without mistakes. 2. Answers will vary but may include other scientists could not repeat the experiment and might not accept the results. **Day 2:** 4 (observation), 2 (hypothesis), 5 (comparison), 7 (presentation), 1 (problem), 3 (experimentation), 6 (conclusion); **Day 3:** 1. experimentation; 2. hypothesis; 3. presentation; 4. conclusion; **Day 4:** 1. C; 2. B

Page 20
1. Answers will vary.

Page 21
Day 1: 1. curious, watchful, patient, eager, creative, persistent; 2. Answers will vary. **Day 2:** 1. Answers will vary but may include scientists must be able to share their findings so that non-scientists can learn and other scientists can use the information and build on it in their work. 2. Answers will vary but may include they can write reports and speak at meetings. **Day 3:** 1. Answers will vary but may include he was eager to learn about sound, and he was persistent. **Day 4:** 1. Answers will vary but may include she was watchful, and she communicated information she learned.

Page 22
1–2. Answers will vary.

Page 23
Day 1: 1. Matter is anything that has mass and takes up space. 2. A solid has a set shape and volume. A liquid takes the shape of its container, and it has a set volume. A gas does not have a set shape or volume. It expands to fill its container. **Day 2:** 1. Atoms are the smallest pieces of matter and cannot be seen. 2. gas; 3. solid; 4. liquid; **Day 3:** 1. P: paint, make a mixture, rip, fold, write on; C: cook, rust, burn, fruit browning; **Day 4:** 1. Water is a liquid. When enough heat is taken away, it changes to ice, which is a solid. When enough heat is added to water, water changes to vapor, which is a gas. 2. 15 minutes

Answer Key

Page 24
1. desk, solid; orange juice, liquid; air, gas; water vapor, gas; book, solid; hot chocolate, liquid; 2. A chair is a solid. The atoms are packed tightly together, so they cannot move. 3. A potato has more mass than a potato chip because it has more atoms, so it is bigger and heavier. 4. When the flour, sugar, eggs, and other ingredients are mixed together, it is a physical change. When heat is added and the cookies bake, it is a chemical change. 5. Iron in the shovel interacts with oxygen and water to make a chemical change called rust.

Page 25
Day 1: 1. A force is a push or pull. 2. A foot kicks the ball and forces it to move forward. The ball stops with the force of a foot or hand on it, or with the force of friction from the grass. **Day 2:** 1. Gravity is the force that pulls two objects toward each other. 2. It is harder to walk up the stairs because you are working against the force of gravity. It is easier to walk down the stairs because gravity is pulling you down.
Day 3: 1. A. television; B. book; C. refrigerator; 2. The items have more mass, so it takes more force to move them.
Day 4: 1. friction; 2. lubricant; 3. inertia; 4. speed

Page 26
1. You will see an object's movement change. 2. Both objects will hit the ground at the same time because the pull of gravity is the same. 3. It would be easier to ride on cement because there would be more friction. 4. B; 5. Yuri applied more force and had more speed. 6. 13 minutes

Page 27
Day 1: 1. wedge; 2. wheel and axle; 3. inclined plane; 4. screw; 5. lever; 6. gear; 7. pulley; **Day 2:** 1. lever; 2. inclined plane; 3. wedge; 4. pulley; 5. screw; 6. wheel and axle; 7. inclined plane; 8. lever; **Day 3:** 1. Answers will vary. **Day 4:** 1. Beth can use the screwdriver as a wheel and axle by putting the tip of the tool into the opening and twisting her wrist to lift the lid. She can also use the screwdriver as a lever to push the lid straight up.

Page 28
1. B; 2. D; 3. Shannon did not do work because the door did not move. 4. You can use less force to do the same amount of work with most simple machines. Simple machines also change the direction of the force. Some can do both. 5. A wheelbarrow has a wheel and axle that helps roll a load with less force. It also has a lever, which changes the direction of the force so that a heavy load can be lifted more easily.

Page 29
Day 1: 1. straight; 2. reflected; 3. refracted; **Day 2:** 1. A rainbow forms when sunlight enters a raindrop. The light bends, or refracts, as it enters the water. Since visible light travels at different speeds, they bend at different angles and the colors separate. **Day 3:** 1. red, orange, yellow, green, blue, indigo, violet; 2. An apple looks red because it absorbs all of the other colors of white light. Only the red light is reflected back to our eyes, so we are able to see a red apple. **Day 4:** 1. Something is transparent if it is clear and easy to see through. Objects will vary but may include a window. 2. Something is translucent if some light passes through it., but it is not clear. Objects will vary but may include a veil. 3. Something is opaque if no light can pass through the object. Objects will vary but may include a wall.

Page 30
1. A triangular prism is a three-dimensional shape. It has three faces that are rectangles and two faces that are triangles, nine edges, and six corners. 2. You will see all of the colors except yellow. The yellow paper is reflecting the yellow color back. It is absorbing the other colors in the spectrum. 3. Light moves through different kinds of matter at different speeds. When the light leaves the air and enters the water at an angle, it changes direction and speed. It makes the handle of the net look broken. 4. The water acts like a mirror. Light is reflected off of the surface. 5. A

Page 31
Day 1: 1. lightning, electricity, burning candle, cooking food, sun, fireworks; 2. Heat is when thermal energy moves from one place to another. **Day 2:** 1. Answers will vary but may include: a stove cooks food. The heat of electricity moves to the food and heats it. A hairdryer uses electricity. It heats the air, which dries my hair. Water for my shower gets hot from electricity or gas. **Day 3:** 1. When something is hot, the particles are moving quickly. If the hot particles touch something that is cooler, the particles begin to bump into each other. The hotter particles share their energy, causing the slower particles to get hot and move more quickly. **Day 4:** 1. A conductor is a material that can move thermal energy easily. Examples will vary but may include metals. 2. An insulator is a material that will not allow thermal energy to move easily. Examples will vary but may include rubber, plastic, or wood.

Page 32
1. The particles in the fudge sauce are hot. They move quickly and bump into the cooler particles in the ice cream and share their energy. As a result, the particles in the ice cream begin to get hot and move faster, which causes the ice cream to melt. 2. Daysha should buy a wool coat. Wool is a better insulator and will help to trap her body heat to help her stay warm. 3. A frying pan is made of metal and is meant to transfer heat to food, so it is a conductor. 4. true; 5. true; 6. false; 7. false

Answer Key

Page 33
Day 1: 1. light; Examples will vary but may include a lamp.
2. sound; Examples will vary but may include a music player.
3. heat; Examples will vary but may include an oven.
4. movement; Examples will vary but may include a fan.
Day 2: 1. electron, proton, neutron; 2. electron; 3. negative;
Day 3: 1. One part of the circuit is open and incomplete, so an electric current cannot flow through it. 2. 15 lightning strikes;
Day 4: 1. Answers will vary but may include do not go near electric power lines that are on the ground. Keep electric appliances away from water. Do not stick things into outlets.

Page 34
1. Check students' raps.

Page 35
Day 1: 1. a rock known for its magnetic properties;
Day 2: 1. compass; 2. north; 3. attract; 4. repel; 5. poles;
Day 3: 1. Check students' drawings and that the magnetized needle and directions are correctly labeled. **Day 4:** 1. Answers will vary but may include hanging papers on a refrigerator.

Page 36
1. Answers will vary.

Page 37
Day 1: 1. cells; 2. energy; **Day 2:** 1. Answers will vary but may include: composed of cells, grow, use energy, respond and adapt to their environment, change, and reproduce.
Day 3: 1. cells; 2. adapt; 3. change; 4. reproduce; 5. develop; 6. energy; **Day 4:** 1. An icicle is a nonliving thing although it can grow longer, since it is not made of cells. 2. A leaf dies when it no longer receives proper light, water, and nutrients.

Page 38
1. Check students' charts to verify characteristics of living and nonliving things.

Page 39
Day 1: 1. trait; 2. species; 3. inherit; 4. diversity; 5. heredity;
Day 2: 1. Answers will vary but may include it may be stronger in one characteristic, which gives the animal a better chance to survive. A species that is different is more interesting. Differences may mean that an animal will reproduce more and pass on those stronger traits. 3. Answers will vary but may include being different may make an animal more noticeable and cause it to be an easy prey. An animal that is different may not be accepted by the species and may be left to die.
Day 3: 1. Answers will vary but may include they need to inherit traits so that they can survive in the habitat. 2. Answers will vary. **Day 4:** 1. Answers will vary but may include snakes inherit their skin color, which helps camouflage them. 2. Answers will vary but may include polar bears inherit their thick fur, which helps protect them from the cold. 3. Answers will vary but may include camels inherit the ability to go without water for long periods which helps them survive in the desert. 4. Answers will vary but may include fish inherit the shape of their fins, which helps determine if they swim fast or slow.

Page 40
1. Heredity makes sure that certain traits are passed down in a species so that the animals can survive in their habitat.
2. Answers will vary but may include differences affected by body systems, such as a lion growing larger than other lions of the same age. Also, differences in inherited traits may occur due to talent. For example, a frog may have better aim and catch more bugs than another frog. 3. Answers will vary.

Page 41
Day 1: 1. Plants need water, air, sunlight, and soil. 2. $4\frac{1}{2}$ liters of water; **Day 2:** 1. A. flower; B. leaf; C. stem; D. roots; The flower makes seeds. The stem carries water from the roots to the rest of the plant and holds the plant up. The leaves take in sunlight and air that makes the food for the plant. The roots hold the plant in the soil and take in the minerals from the soil and water. **Day 3:** 1. A seed is the part of the plant from which a new plant can grow. 2. Answers will vary but may include some plants push the seeds away. Water and wind can carry seeds to new places. Animals eat some seeds. The seeds move through their bodies and may be dropped far away. Animals can carry away the seeds and bury them in the ground as a way to store food. Seeds also stick to people's clothing or animals' fur as they move past plants. The seeds fall off somewhere else. **Day 4:** 1. true; 2. false; Chlorophyll is a green pigment that helps plants absorb sunlight. 3. true; 4. true; 5. false; Plants give off oxygen as waste.

Page 42
1. Anna will see an area the same size of the circle that is yellow or less green than the rest of the leaf because that part of the leaf was not able to absorb sunlight. 2. Plants need light, chlorophyll, carbon dioxide, and water to make food. The process is called photosynthesis. 3. Answers will vary but may include they provide oxygen, food, shade, lumber to build with, and shelter for animals.

Answer Key

Page 43
Day 1: 1. Animals need food, air, water, and shelter. 2. Answers will vary. **Day 2:** 1. Answers will vary but may include a giraffe has a long neck to help it reach the leaves that are in the tallest trees. **Day 3:** 1. Method 1: They run. Examples will vary. Method 2: They hide. Examples will vary. Method 3: They use body parts to defend themselves. Examples will vary.
Day 4: 1. migrate; goose; 2. adapt; arctic hare; 3. hibernate; bear

Page 44
1. Fish have gills that help them filter oxygen out of the water. 2. The desert kangaroo rat gets water from the seeds it eats. 3. Animals need shelter to stay safe from predators and to protect themselves from the weather. 4. Answers will vary but may include the hawk has good eyesight to see animals from high up in the air. It uses its wings to fly down to where the food is. It has a sharp beak to kill small animals and tear them apart. It has sharp claws to catch animals. 5. Answers will vary but may include trees are being cut down in forest habitats, pond habitats are being filled with dirt and rocks, and people may grow lawns and gardens in desert habitats.

Page 45
Day 1: 1. animals with backbones (vertebrates) and animals without backbones (invertebrates); 2. Insects do not have backbones, so they are invertebrates. 3. Answers will vary. **Day 2:** 1. Answers will vary but may include reptiles are covered in scales and are cold-blooded. They use lungs to breathe. Examples may include alligators and snakes. 2. Answers will vary but may include amphibians are cold-blooded animals. They lay eggs in water. They breathe with gills when they are young, but grow lungs as adults. They spend part of their life in water and part on land. Examples may include turtles and frogs. **Day 3:** 1. Answers will vary but may include birds breathe with lungs. They are warm-blooded. They have wings and are covered in feathers. Birds make nests and lay eggs. Examples may include penguins and robins. 2. Answers will vary but may include fish have scales and live underwater. They lay eggs. Examples may include trout and salmon. **Day 4:** 1. Answers will vary but may include mammals breathe with lungs. They are warm-blooded. They have fur or hair. Female mammals give birth to live young and feed their young with milk from their bodies. 2. mammals; Answers will vary but may include whales give birth to live offspring, feed their young with milk from their bodies, have backbones, breathe with lungs, and are warm-blooded.

Page 46
1. The two groups are animals with backbones and animals without backbones. The backbone is part of the skeleton inside the body of the animals. Animals that do not have backbones often have a hard outer skeleton to keep the inside parts safe. 2. It is an insect or an invertebrate. It has three body parts and six legs. It has a hard outer shell. 3. A bat is a mammal. It has wings like a bird, but the female feeds its young milk from its body. 4. They have other characteristics that put them in different groups. A bird has feathers and a backbone. A butterfly has scales on its wings and an outer shell on its body. 5. Check students' poems.

Page 47
Day 1: 1. A life cycle is the sequence of steps that a living organism follows from birth to death. 2. Not all living things look like the adult. A plant begins life as a seed and grows into something that has leaves and a stem or trunk. A frog begins as an egg and grows into a tadpole and then a frog. A dog and baby look very much like the adults. **Day 2:** 1. seed; 2. root; 3. seedling; 4. stem, leaves; 5. flowers, cones; 6. life cycle; **Day 3:** 1. Metamorphosis is a major developmental change some animals experience as they grow from egg to adult. 2. 2, 4, 1, 5, 3; **Day 4:** 1. Check students' graphs.

Page 48
1. A. egg; B. caterpillar; C. pupa; D. adult butterfly; 2. A female butterfly lays many eggs on the leaves of plants. The eggs hatch into caterpillars. The caterpillars eat lots of leaves. Then, each caterpillar forms a hard shell and rests. This is the pupa stage. Soon, the insect breaks out of the case. It is an adult butterfly that has wings, six legs, and two antennae.

Page 49
Day 1: 1. all of the living and nonliving things in a place; 2. the place an animal lives where all of its needs are met; 3. a group of one kind of living thing living in a place; 4. everything that is around a living thing; 5. all of the groups of living things living in a place; 6. Answers will vary but may include students, teacher, books, desks, rulers, or pencils. **Day 2:** 1. Answers will vary but may include squirrels, trees, birds, moss, and ants. 2. Answers will vary but may include rocks, air, sunlight, soil, and water. **Day 3:** 1. desert; 2. rain forest; 3. arctic or tundra; 4. ocean; 5. pond; **Day 4:** 1. Answers will vary but may include: flood, fire, drought, disease, and the building of roads, buildings, and farms. 2. Answers will vary.

Page 50
1. Check students' drawing and labeling.

Answer Key

Page 51
Day 1: 1. A producer is a living thing that makes its own food. 2. A producer gets its energy from sunlight and soil. 3. Plants are important producers because they provide food for many animals that eat them directly. They also provide food indirectly for animals that eat the plant eaters. **Day 2:** 1. Herbivores eat only plants. Examples will vary but may include horses. Carnivores eat only meat. Examples will vary but may include polar bears. Omnivores eat both plants and meat. Examples will vary but may include humans. 2. Consumers get their energy from the food they eat. **Day 3:** 1. A decomposer is a living thing that gets energy by breaking down dead things or wastes that living things leave after digestion. Examples will vary but may include fungus and worms. 2. Decomposers are important because they can access stored energy in dead things and wastes so that it can be used. **Day 4:** 1. A food chain shows the flow of energy. 2. 3, 1, 4, 2

Page 52
1. C: insect, human, bear; P: lettuce, rose, tree; D: mushroom, worm, bacteria; 2. The food chain might be found in a forest. A plant is the producer. A rat is a consumer that eats the plant to get energy. The snake eats the rat to get energy. Finally, the hawk is the consumer that eats the snake to get energy. 3. A

Page 53
Day 1: 1. A mineral is a nonliving thing that comes from Earth. 2. Answers will vary but may include hardness, color, and shape. 3. Rocks are made out of minerals.
Day 2: 1. Type 1: Sedimentary rocks form when many layers of material pile on top of each other. They are pressed together and harden. Type 2: Igneous rocks form when hot, liquid rock cools and hardens. Type 3: Metamorphic rock is made from sedimentary or igneous rock that has been heated under pressure. **Day 3:** 1. sedimentary; Sedimentary rocks are soft rocks and can be easily broken. 2. metamorphic; It had different kinds of minerals in it. 3. igneous; Hawaii is an island formed by volcanoes. As the hot lava cools, it turns black. The holes are from air bubbles. **Day 4:** 1. true; 2. true; 3. false; 4. true; 5. false

Page 54
1. A. igneous; B. sedimentary; C. metamorphic, 2. Each shelf has 9 rocks. 3. The sculptor will use marble, a metamorphic rock, because it is harder rock than limestone, which is a sedimentary rock.

Page 55
Day 1: 1. the shape of a living thing left as an imprint; 2. the study of fossils; 3. something left over from a living thing that died long ago; 4. the shape of a living thing made when mud or minerals fill a space; 5. a kind of animal that lived long ago; **Day 2:** 1. 3, 2, 1, 5, 4; **Day 3:** 1. There are fossils of organism parts, imprints of tracks, and casts of their bodies.
2. A paleontologist must work carefully so the fossils are not lost or harmed. The better and bigger the fossil, the more a scientist can learn. **Day 4:** 1. Answers will vary but may include the tooth that is sharp came from an animal that ate meat, since sharp teeth are needed to tear meat. The tooth that is flat came from an animal that ate plants, since flat teeth help to chew plants.

Page 56
1. Sedimentary rock is made of small pieces of rocks, sand, and mud. The layers of sediment cover and preserve the animal or plant. The small pieces of rock take the shape of something and give more details. The heat used to form igneous and metamorphic rocks would destroy specimens. 2. In a mold, the parts of the organism were dissolved and left an imprint. In a cast, rock and mud fill the area that was dissolved to make a three-dimensional shape. 3. Scientists can compare what they find with animals living today. If the parts are the same, they can make inferences and draw conclusions that the animals looked and behaved the same way. 4. Answers will vary but may include a brush, chisel, and drills.

Page 57
Day 1: 1. a very high, pointed piece of land; 2. a low place between mountains; 3. a deep valley with steep, high sides; 4. a wide, flat area of land; 5. flat land rising above the surrounding land; 6. a piece of land totally surrounded by water; **Day 2:** 1. Weathering is a process, either chemical or mechanical, where rocks are broken down into smaller pieces. 2. Weathering is caused by water, wind, and ice. 3. Roots grow in the cracks of rocks. As they get bigger, they break the rocks. **Day 3:** 1. Weathering is breaking down the rocks. Erosion is moving the small pieces of rocks to a new place. 2. Answers will vary but may include moving water can pick up pieces of rocks and move them to new places. Glaciers move slowly across the land. They push and pull rocks as they move. Gravity can cause rockslides, which pull rocks down a steep slope. Wind can blow pieces of sand and dust to new places.
Day 4: 1. volcano eruption; 2. earthquake; 3. flood

Page 58
1. B; Answers will vary but may include the rows will slow the water as it flows down the hill and reduce soil erosion.
2. 59 rows; 3. C; 4. A; 5. D

Answer Key

Page 59
Day 1: 1. A; 2. C; **Day 2:** 1. cirrus; 2. cumulonimbus; 3. stratus; 4. cumulus; **Day 3:** 1. A thermometer measures the temperature of a place. 2. An anemometer measures wind speed. 3. A rain gauge measures how much precipitation has fallen. 4. A weather map shows what the weather will probably be like, including temperatures, air flow, and precipitation, for a large area. **Day 4:** 1. Meteorologists are scientists who study the weather. 2. Answers will vary but may include air masses are large bodies of air that can be warm or cold. They move across the land and greatly change the weather, including wind direction and speed, temperatures, and precipitation.

Page 60
1. Answers will vary but may include it affects the clothes I wear and what activities I do. 2. Ben saw 7 lightning bolts. 3. The weather is fair and warm. 4. D; 5. C

Page 61
Day 1: 1. Answers will vary but may include ocean, river, stream, brook, pond, lake, gulf, or tributary. 2. It looks blue because the surface of Earth is mostly covered in water. **Day 2:** 1. 100 liters; 2. 8 liters; 3. 10 milliliters; **Day 3:** 1. Freshwater is found in streams and lakes. People and animals drink freshwater. Salt water is found in the oceans. Many animals we eat live in this water. **Day 4:** 1. the process where water is removed from the surface of Earth and returned back to Earth; 2. when heat is added to liquid water, changing it to a gas; 3. when heat is removed from a gas, changing it to liquid water; 4. water when it is a gas

Page 62
1. It shows the water cycle. 2. The sun heats water on Earth and changes it to water vapor through evaporation. The water vapor rises into the air. The vapor cools as it rises through condensation and changes back into tiny droplets of water. The tiny droplets of water get packed closely together and form clouds. The drops become heavy and fall as precipitation, which run off and collect in bodies of water. 3. A; 4. B

Page 63
Day 1: 1. Mercury, Venus, Earth, Mars, Jupiter, Saturn, Uranus, Neptune; **Day 2:** 1. sun; 2. gravity; 3. orbit or revolve; 4. rotates; 5. asteroid; **Day 3:** 1. true; 2. false; 3. true; 4. false; 5. true; **Day 4:** 1. The inner planets are Mercury, Venus, Earth, and Mars. Answers will vary but may include they are alike because they are closer to the sun, they have rocky surfaces, are warmer, and are smaller. 2. The outer planets are Jupiter, Saturn, Uranus, and Neptune. Answers will vary but may include they are colder, they have more moons or rings, they are made mostly of gas, and they are larger.

Page 64
1. The two materials that support life are water and an atmosphere with oxygen. 2. Planets that are closer to the sun are hotter and orbit more quickly. 3. Answers will vary but may include everything on Earth would die. Living things need the sun. The sun allows plants to grow, which are used for food. The sun is an important part of the water cycle that provides living things with fresh water. 4. C; 5. D

Page 65
Day 1: 1. true; 2. false; The moon reflects light from the sun. 3. false; The moon is the largest object in the night sky. 4. true; **Day 2:** 1. rotates; 2. twenty-four; 3. night; 4. day; 5. twenty-nine; **Day 3:** 1. The moon orbits around the sun. As it moves, the sun lights different amounts of its surface. 2. During a full moon, the moon is shown as a complete circle. The dark half faces away from the Earth. 3. During a new moon, the moon cannot be seen at night. The lighted half of the moon faces away from Earth. **Day 4:** 1. Earth tilts on its axis, and it revolves around the sun. 2. It is summer.

Page 66
1. It is winter. 2. The days are shorter because there is less light shining on this part of Earth. The weather is cooler because the sun's rays are angled and not as direct. 3. There would not be any seasons. 4. C; 5. D

Page 67
Day 1: 1. Answers will vary but may include scientists might be concerned because animals and plants that we do not even know about are dying. They might be important to the balance of the rain forest. **Day 2:** 1. Answers will vary but may include it is not a good idea to leave the land bare. Without the plants, wind and water will cause erosion and remove the soil. The land is flat and will easily flood too. **Day 3:** 1. Smoke from the fires cause pollution in other parts of the world. Native people continue to clear land, which leaves land bare, resulting in soil erosion. **Day 4:** 1. The drug companies are finding new medicines to help people. The native people are earning money, so they want to protect the forest.

Page 68
1. C; 2. A; 3. Answers will vary. Check students' writing.

Page 69
Day 1: 1. Paper is thrown out the most. 2. Answers will vary but may include boxes, office paper, package wrapping, and paper towels. 3. Paper and Other; **Day 2:** 1. something that is thrown out; 2. to use less of something; 3. to find a new use for something; 4. to make something new out of something old; 5. to protect something from destruction; **Day 3:** 1–2. Answers will vary. **Day 4:** 1. water; 2. electricity; 3. clothes or shoes; 4. paper; 5. aluminum

Page 70
1. C; 2. D; 3. Answers will vary. 4. Answers will vary but may include some minerals, like aluminum, are nonrenewable. They will run out some day. By reusing, reducing, or recycling, we can make our limited resources last longer.

Answer Key

Page 71
Day 1: 1. A resource is something in nature that living things use. 2. air, cow, corn, oil, water, coal; **Day 2:** 1. Answers will vary but may include a rock is a resource because people use it to build things, like houses, bridges, and walls. A chicken is a resource because it can be used for food, it lays eggs that can be used for food, and its feathers can be used for pillows. **Day 3:** 1. A renewable resource is a resource that can be replaced in a person's lifetime. Examples will vary but may include trees, crops, and animals. 2. An inexhaustible resource is a resource that can be used repeatedly and not be used up. Examples will vary but may include sunlight, air, and water. **Day 4:** 1. One-half of the trees were cut down. 2. One-tenth of the trees were sold to the paper company.

Page 72
1. Answers will vary but may include a tree is a resource because people use it for many things, like paper, lumber, heating, cooking, and furniture. 2. Yes, it is important to care for inexhaustible resources because we need these things to live. Water is an inexhaustible resource, but if we do not keep water clean, we cannot drink it and plants cannot use it to help them grow. 3. A lumber company can plant new trees to replace ones they cut down. 4. C; 5. D

Page 73
Day 1: 1. to keep the living things on Earth safe; 2. every one of this kind of living thing has died; 3. there is only a small number of this living thing; 4. a group of living things that is being kept safe by laws; 5. a place where the habitat of a living thing is kept safe; **Day 2:** 1. There are fossils of bones, teeth, and prints left behind in rocks. 2. Answers will vary but may include changes in nature, such as climate and land changes caused the dinosaurs to die. **Day 3:** 1. The bluebird did not have the proper shelter. 2. People changed the land and removed the bluebird's habitat. **Day 4:** Answers will vary but may include people put the bluebird on the endangered list so that it would be kept safe by laws. People built boxes that the birds could use to make nests and placed the boxes in habitats where the birds had everything they needed to live.

Page 74
1. People sprayed a harmful chemical. 2. The chemicals were not used near eagles, but scientists realized the birds were getting the chemical in a food source. 3. Answers will vary but may include plants and animals are all connected. If something harms one living thing, it often affects other living things. Animals or plants might die as a result. It is important to realize the connections to keep nature safe and balanced.

Page 75
Day 1: 1. detectors; 2. escape; 3. outside; 4. extinguisher; 5. clothes; **Day 2:** 1. true; 2. false; Do not stand under a tree if it is lightning. 3. true; 4. true; 5. false; Do not talk on a landline phone if it is lightning. **Day 3:** 1. Answers will vary but may include move under a sturdy table or crouch down and tuck your head. Stay away from glass windows, doors, and heavy objects that might fall down. 2. Answers will vary but may include move to an open area away from buildings, gas lines, and power lines. **Day 4:** 1. tornado; 2. glass windows; 3. ditch or low-lying area; 4. reports; 5. over or passed

Page 76
1. Glass can break and fly around. It might cut someone. 2. Lightning is attracted to tall objects. A tree is a likely target during a thunderstorm because it is tall. 3. Smoke detectors sense smoke and beep loudly. Even if someone is asleep, the loud noise will alert people that there is smoke nearby. 4. Answers will vary but may include people need to learn how to stay safe in dangerous situations. Practicing safety rules helps people know what to do in a dangerous situation. If they practice, they are more likely to follow the steps correctly. 5. Answers will vary but may include the number 911 goes directly to emergency services. It is an emergency number that is to be dialed when there is a dangerous, life-threatening situation and help needs to be sent immediately. Examples will vary.

Page 77
Day 1: 1. grains, vegetables, fruits, dairy, and proteins; Examples will vary. **Day 2:** 1. Answers will vary. **Day 3:** 1. A balanced diet is eating the right amount of foods from each food group. 2. 15 grams; **Day 4:** 1. protein; 2. vitamin A; 3. calcium; 4. iron; 5. carbohydrate; 6. vitamin C

Page 78
1. Answers will vary but may include a bigger body needs more food and nutrients than a smaller body. Some people are more active or in a growing phase and need more of one food and nutrient to power the body. 2. All foods have different nutrients. The body needs a variety of nutrients to be healthy. 3. Check students' reports. 4. Answers will vary. 5. D

Page 79
Day 1: 1. Exercise makes the body strong, helps control weight, helps control stress, improves sleep, and prevents illness. 2. The muscles get smaller and cannot work as well. Muscles can be more easily injured when doing activities, falling, or moving suddenly. **Day 2:** 1. The heart beats faster, the breathing rate increases, and blood flows more quickly through the body due to increases in temperature. 2. Answers will vary but may include: It is easier to move during the activity. There is less chance of being injured. **Day 3:** 1. running drills; 2. 44 minutes; **Day 4:** 1. It gives the body a chance to return to its normal rhythms. 2. It helps keep the muscles from being sore and stiff.

Page 80
1. false; 2. true; 3. true; 4. true; 5. true; 6. Different activities work different muscles, which makes sure the whole body gets exercise. 7. Answers will vary but may include Jana did not warm up or cool down. She might have sore muscles later. 8. A; 9. D

Page 81
Day 1: 1. The wind is more constant and more forceful farther above Earth's surface. 2. 180 times; **Day 2:** 1. Not all power can come from water because there are many places on Earth that do not have access to lots of water. **Day 3:** 1. The color black absorbs the most light and better heats the water. **Day 4:** 1. Fuels that are used in many power plants may run out. Since atoms are the main source of energy in nuclear energy, they will not run out.

Page 82
1. Answers will vary but may include wind is a good source of energy because it is an inexhaustible resource and cannot be used up. However, wind might not blow all of the time, which might keep the generators from producing electricity when people are depending on it. 2. Answers will vary but may include a homeowner living in Florida would be more likely to use solar power because the sun shines more often there. The source of heat would be constant. Alaska has periods of little light, so the water would not stay hot. Also, snow might cover the panels, so no light could be collected. 3. Answers will vary but may include people use the lake behind the dam for fun activities, such as boating, swimming, and fishing. Dams provide a habitat for water plants and animals to live. 4. Answers will vary.

Page 83
Day 1: 1. A computer is a device that can store information and work to solve problems very quickly. 2. An abacus was used to add and subtract. It solved problems very quickly. **Day 2:** 1. the part where people see the information; 2. the part on which people input information; 3. the part that does the computing; 4. the unit in which information is stored; 5. the network that links computers all over the world; 6. a program that makes the computer work; 7. the part that stores all the information; **Day 3:** 1. Answers will vary but may include email, blogs, personal websites, cameras, and microphones. **Day 4:** 1. Answers will vary but may include it stores information. It has a screen to view pictures. It has a microprocessor and memory. It needs software to make it work.

Page 84
1. Answers will vary. Check students' writing.

Page 85
Day 1: 1. She won a famous prize two times, and she discovered two new elements. She also developed the use of X-rays, which greatly helped people. **Day 2:** 1. A satellite allows people to communicate all around the world. A doctor in a faraway country can ask for help and information from a doctor who has more information. **Day 3:** 1. A classification system helps scientists tell how things are alike and different. Classifying stars helps put them into groups so that scientists can learn more details about each group. **Day 4:** 1. Potter's pictures had many details. Other scientists could look at the pictures and compare the samples to fungi they were studying to see how they were alike and different.

Page 86
1. a scientist who studies stars; 2. a scientist who studies matter and energy; 3. a scientist who studies materials and how they work together when joined; 4. a scientist who studies plants; 5. Answers will vary. 6. Answers will vary but may include scientists work to learn new ideas. Many times, the ideas can help people. They share the information they learn. Examples will vary. 7. A; 8. C

Page 87
Day 1: 1. The plants added nutrients to the soil. **Day 2:** 1. A scientist would be able to protect his idea and use it to make money. **Day 3:** 1. Edison wanted to link the electricity in houses and businesses to make it possible and affordable for everyone to have electric lights. **Day 4:** 1. The bulb was safer because the screw bottom held the bulb in the lamp and kept it from falling out and breaking.

Page 88
1. true; 2. true; 3. true; 4. true; 5. true; 6. Edison shared his work with people. Latimer studied Edison's work and found a way to make the lightbulb better. 7. Edison had a good device, but it was not useful unless many people could benefit from using it. He needed to find a way to get his invention to many people. Since there was no way yet invented, Edison had to create one. It meant solving many different problems, which led to a system for lighting homes and businesses. 8. Answers will vary but may include the peanut crops grew very well. Carver had to find ways to use the crops so that people could make money and continue to grow the plants.